Dark Web

Everything You Didn't Know About
Dark Web

*(Know All Everything You Need About
Exploiting the Dark Web)*

Cynthia Slaton

Published By   **Jordan Levy**

Cynthia Slaton

All Rights Reserved

Dark Web: Everything You Didn't Know About Dark Web (Know All Everything You Need About Exploiting the Dark Web)

ISBN   978-1-7780579-3-9

No part of this guidebook shall be reproduced in any form without permission in writing from the publisher except in the case of brief quotations embodied in critical articles or reviews.

Legal & Disclaimer

The information contained in this ebook is not designed to replace or take the place of any form of medicine or professional medical advice. The information in this ebook has been provided for educational & entertainment purposes only.

The information contained in this book has been compiled from sources deemed reliable, and it is accurate to the best of the Author's knowledge; however, the Author cannot guarantee its accuracy and validity and cannot be held liable for any errors or omissions. Changes are periodically made to this book. You must consult your doctor or get professional medical advice before using any of the suggested remedies, techniques, or information in this book.

TABLE OF CONTENTS

Introduction

Do you recognize this?

If you've ever heard shocking reports about illegal online drug stores, hit men hired, famous people arrested on child porn charges, crazed experiments in science and Illuminati rituals, then you've likely heard about "the "dark web" or the deep web.

It's believed to be an uncharted web surfing experience, the enigmatic and often frightening "dark part" on the Internet that is where you'll be sure to find items that are shocking, illegal , or highly classified.

This is a fantastic story for news magazines with a sensationalist angle to take on, particularly when you've got unconfirmed stories of cults, aliens murders, and other shocking events that no reasonable human being would ever want to see. It's also a top choice site on YouTube Horror and CreepyPasta as they are fond of making urban legends more interesting.

Have you ever thought whether these tales are true? What exactly is this Deep Web or Dark Web in particular?

It's foolish to claim that all the stories you've read regarding the Deep Web are fake. Certain cases have been tragically real and often widely reported. There was a real online drug-dealing website called Silk Road which was shut off in the hands of FBI. There have been numerous cases of murderers and pedophiles being caught through the investigation of the IP address of deep websites as well as "digital fingerprints".

There have been also well known stories about national security secrets and whistleblowing that have been revealed by the likes by Edward Snowden, Bradley Manning Julian Assange, Bradley Manning, the "Anonymous" crowd of demonstrators, and many more.

What we can say that is for certain is that there's plenty happening in the world of the deep web with some that may be frightening. We also recognize that curiosity eventually going to be the king of you and you're likely be compelled to explore this forbidden area by yourself.

We're here for you to let you know that, shockingly, it's totally safe and legal to use through the Deep / Dark Web and it's very unlikely to be snatched, escorted into the Red Room, or arrested in the mishap of stumbling across the wrong web site.

It is likely that if you are taking the necessary precautions and in addition, understand exactly how darknets and the deep web function, you will be less likely to commit a mistake.

What we're trying to do within this book is teach you how to browse anonymously, hide your tracks in the event that you're a bit suspicious, and, generally you can avoid getting caught for doing something irrational, which we believe you're not likely to do!

At the end of this book you'll know what software you'll need and what additional measures you need to take, and what websites could pose the greatest personal risk, and ought to be avoided.

So relax and enjoy while we embark to the most frightening internet on the planet--the unfiltered, undiscovered Internet!

Chapter 1: What's The Deep Web And Why Is It

To Be Explored?

Words such as "deep web" or "dark web" are often used interchangeably , even though they are two distinct entities. The simplest explanation as to that dark and deep can be confused is that the vast majority of internet users are only using "the "surface web"--the most well-known and frequently hyperlinked websites available in the Internet. There are millions of privately-owned and managed websites that no one has yet officially identified as "indexed" (such like Google or Bing perform by crawling publicly-owned websites) or perhaps one or two people actually have seen.

Some have compared the dark and deep web to fishing deep in the ocean. Or an iceberg that appears up to the surface, but is thousands of miles beneath the ocean. In 2001, the dark web was believed to be "orders of magnitude" greater than the surface web and the latest reports from 2016 suggest that hidden databases and websites are at least 500 times more in the number of sites than that we can see on the surface web.

In theory the internet is full of millions of websites normally you wouldn't find except if you could get into the dark and deep web.

What we see on our most visited websites (Face book, YouTube, etc.) could be referred to as the filtering of media. The media you can access will be only news and information that has been filtered information, along with entertainment websites, that have been carefully chosen to provide you with the best viewing experience. It's similar as how the executives design their cable or network television.

Search engines like Google and Bing could be able to connect you to websites that are not gaining much popularity, due to a longtail search like "Historical records dating back to the 18th century." It might not be a popular website, but you'll be able to locate it with a bit of effort.

Sites such as LiveLeak as well as 4Chan are well-known for posting content that has been deemed controversial, and is later scrutinized and removed However, these sites remain considered to be the top of the web due to their an enormous amount of usage (and lots of people who are able to complain about content that is illegal or questionable) and also because they have a lot of pages that are searchable.

The term "deep web" refers to pages and sites that are deliberately excluded from search engines that index the contents. Sometimes referred to as the invisible web or hidden web These sites and pages generally aren't illegal or dangerous, they're just not designed to be indexed by search engines. It could be everything from entertainment that is not public, or even private content that's secured with a password, for example, internet banking records, online mail, pay-on-demand video or subscription magazines as well as other legal or medical documents which are considered to be not public information.

It is worth noting the two options to prevent being crawled by search engines You can either install the code manually (or use an auto-installer) which blocks search engines from crawling your website or connect the website (or the site) to other websites.

Search engine robots discover new websites by following the link trail, from pages to pages, onsite and offsite. A website that does not advertise and does not sign up with any public search website or domain name, it can still be found through search, if it's connected to other websites.

The pages of deep web cannot be searched, and therefore offer more security than websites that are public. However, these sites remain accountable to an authority as these sites are run by well-known internet service providers who provide servers that are shared or private. Although they are private however, everything you see on these sites isn't private. There's a long paper trail. Deep web pages could be restricted by these techniques:

* Registration and login are required.

* Pages that are not linked

Non-HTML content like multimedia that has particular file formats

* Scripted content that can be only accessible via JavaScript, Flash or other specific scripting

" Dynamic Content" or context-based pages that are only accessible via a specific query, form or access context

* Search-prohibited codes like Robots Exclusion programming, or even using CAPTCHAs in order to disqualify search bots

In simpler terms the deep web is theoretically accessible, even although it's not always convenient.

The Differential ties Between Dark Web and the Deep Web and Dark net

What separates the "deep web" in comparison to that of the "dark web" is the fact that it is not accessible to access to the general Internet and isn't accessible without the use of special software. It is impossible to accidentally discover it accidentally. Dark Web. You have to be careful to download the program.

Darknet is a term used to describe an overlay network with limited access and requires the use of different communication protocols and ports to be configured. There are many darknets accessible via"the "Dark Web" that is the other web you're not supposed discover.

Perhaps you've seen software for sharing files such as Napster or LimeWire in the past and these are two examples of a "darknet" type, the peer-to-peer community of users sharing files. This is the "Deep Web" tends to include the darknet. The main difference lies in the actions you need to take in order to join the darknet.

The concept of the "darknet" actually predates it being called the "Deep Web" as this Darknet concept was first proposed in the 1970s and was a reference to networks that were distinct from ARPANET that later evolved into the global connected Internet. This Deep Web resulted because of the need to impede the searchability of the Internet. The Darknet concept was born out of the need for privacy and more private traffic.

The major distinction of Deep as well as Dark is that a darknet isn't equipped with the standard "searchability" as well as the capability to be downloaded by uninitiated users. If you tried loading darknet URLs from underground sources, you'd not be able load them by simply clicking the link on Internet Explorer, Firefox, Safari, Chrome and so on. The darknet is often thought of as part or the Deep Web in technical terms but in the sense of semantics, it's something much more secure.

In the next book, we'll examine the best methods access the Dark Web as well as the most secure methods to use and which ones carry the risk, and which software are best for browsing in anonymity.

The reason is it that the Surface Web is Extremely Limited

Many people ask when they find out that there is a "dark part of the Internet" exists is: what is the reason you would look up information that is considered "secret"?

The most obvious response is that web is very restricted in its ability to provide details, and like how TV networks design their schedule, there's no "exploring" to be done on the internet's surface.

The web's surface is protected and is filtered for general access. Some people from English-speaking nations might believe that Darknet software is only beneficial to people living in communist countries, in which the government is in charge of Internet access, take a look at the facts of capitalist and socialist countries.

Google and Bing in addition to others English search engines have been known to manipulate the results of their keywords to serve esoteric commercial reasons. Some believe that smaller websites are being stifled by larger search engines to promote famous brand name businesses. In addition, many of the news articles you read on the web provide biased information in order to serve different commercial or government purposes.

The excitement when you explore this Dark Web comes from the notion that you might be browsing through stories or websites that offer a totally new view of the world. Be aware that to have a website hosted that is accessible to the public and to be considered a "mainstream" website, the majority of web content that is published must be as "mainstream". Hosting providers will not allow the user to host content that not considered controversial, taboo, or "alternative" in view.

This is why it's no surprise that users are constantly searching darknets for news from the underground in particular, because many whistleblower websites are now blocked from the internet's surface. Some hackers fantasize about becoming one of the "big whistleblower" Imagine what they could discover when they search for "secret" documents which no one else is permitted to be able to access.

Certain users prefer using darknets to safeguard their privacy and to escape what they believe to be "mass monitoring" from intrusive surface websites, as well as the federal government.

In addition, the Darknet can be utilized for illegal activities for instance, providing illegal websites, hacking offenses or sharing illicitly acquired media (new films and music) as well as distributing illicit pornography or even scenes of real-life murder. There's also plenty about counterfeit software as well as identity theft and spam operations.

Then there are those who enjoy surfing the dark web just to find something brand new and exciting. The majority of sites that are found on Darknet are safe and superficial excursions into niche subjects. Many web hosting users do not like being searchable or indexed, and prefer privacy. Some make use of Darknet to access the Darknet market to locate rare books that could not be available by Amazon, BN.com and other traditional retailers.

BBS as well as Usenet Predecessors

In the 1990s and 1980s, Bulletin Board Systems were the precursors to the internet world. They were servers that ran custom software that allowed anyone who had access to a computer access to it with terminal. After login users gained access to a limited group of people. There, they could interact with other users and play games, check out news items and share content with no third-party company overseeing the whole process.

"Systems Operator" of the BBS "Systems Operator" was in charge of the cyber-community with the computer, software and a telephone line that connected users directly on the BBS. Sometimes, the "sysops" who controlled the BBS's the phone numbers of other BBS's you could phone and connect to for more exploration such as file trading, news and sharing.

Utilizing Usenet group was the next leap in technology. They served as repository for discussion groups that connected users from different locations, and permitting uploads and downloads as well as discussion in text. Contrary to BBS or modern Web servers Usenet technology didn't require the use of a central server nor a dedicated administration. It was an information system that forwards news to users and used several servers.

The dark and dark web is essentially a reminiscence of the old Usenet and BBS dayswhen users could access the unregulated, unfiltered, and generally unexplored online world, free of advertisements from companies.

Are you able to surf the Darknet Legal or illegal?

While surfing through the Dark Web, you may uncover something that no one else has seen before. You might uncover things you didn't know about or get an interesting glimpse of a new way of life or a religious or political perspective which you did not know existed.

There is a chance that you will stumble across illegal content, however, there is a way to prevent which we'll talk about in the future. Additionally, you might and likely will find many dead links. It's part of the adventure to unexplored territory.

The excitement of exploring the most dark parts of the web is taking a risk and exploring the unknown. It's perhaps the closest humans have ever come to exploring space, or even the deepest section in the sea. The most exciting part is looking for something you haven't seen or coming on a truly original or mysterious web page.

There is nothing illegal or unsavory regarding exploring the darknet using an open mind. You are entitled to download software to explore these "locked websites" and you must exercise caution whenever you discover something that is illegal. When you're not sure you should abandon the site. For more details on "what could go wrong" make sure you go through our final chapter, which reveals some of the more shocking yet still plausible terrifying stories of deep internet exploration.

Let's move to the first chapter, which examines the best method of surfing securely and anonymously when going to Darknet.

Chapter 2: Best Methods To Surf Anonymously

For more details about the reason to use a pseudonymous browser, check out our subsequent chapters that describe what usually goes wrong when you decide to go to websites with untrustworthy owners. If a user engages in something unlawful, both federal and local authorities might be able to catch them It usually involves people who have given up their privacy when they make one blunder.

Even if the intentions behind surfing aren't harmful but it's not a good idea to share your personal information to users all over the world. And this data about you is kept on your home device(s) as well as in your ISP connections, and holds your name, address, telephone number and your location, in addition to other indicators. It's also quite liberating to browse the internet anonymously free of targeted advertisements and worry about having your personal information stored on your computer.

The issue of the use of "anonymous browsers" in Firefox as well as Chrome "Incognito/Private" can be that it doesn't prevent the history of your browser from saving. Even though the Internet activities aren't stored on your computer does not mean that it's completely anonymous. You're leaving the trace.

Users who are on the web's surface might want to consider eliminating cookies, as they are tracking files that keep details about you. Moreover, the majority of browsers automatically accept cookies. The CCleaner program , you can delete cookies, but you'll require anti-tracking software in order to opt out of these advertising networks that monitor your online habits. (Some popular applications for ensuring privacy on the web are Privacy Badger Ghostery as well as Disconnect.) Many people are unaware that these tiny "cookies" track you by following them across the Internet.

In the end you'll be able to see that just because you're able to clean your device doesn't mean you're being unnoticed by other users. Your actions will still be monitored and monitored from your internet service provider.

There are two essential methods to surf the Darknet for anonymous surfing. The process of connecting to the Darknet requires either specialized software or the configuration of the proxy server.

You need to use a secure anonyme browser to use the web's deep. But, the majority of users employ two ways to protect themselves one of which is the "virtual secure network" which will hide your IP address. This is the identification information you transmit. We'll go over the distinction between these two methods in the next chapter.

If you're interested whether it's possible to browse the darknet with your preferred browser, such as Firefox and Chrome. If you've tried to do this you're probably aware that the browser will issue an error message saying that it is unable to download the contents. This is due to the fact that normal browsers usually have advanced privacy and security settings, along with an inbuilt firewall to shield users from harmful sources. In addition, the numerous web servers that you can access on the internet hidden might also have different privacy and security settings that you need to adjust, to allow access to them.

There are methods to force the normal browser to transform into a dark web browser. It is necessary first download TOR and then set up Chrome and Firefox to function with the new security settings that TOR has altered specifically in the proxy settings.

Many users won't even consider this since it's much easier and safer to just install Tor and then use a Tor-ready web browser that provides privacy-conscious browsing. Utilizing a standard browser and having to always having to tweak the settings, install and switch on private browsing is difficult. There are a few plugins available to Firefox as well as Chrome and, according to reports, block third-party cookies, but do not keep search history or browsing history, and stop all trackers.

However, the best method in order to "go dark" while surfing Darknet is to install and then use the TOR network and browser. This is the topic of the next chapter.

Chapter 3: The Pros And Pros And Tor

T.O.R. is The Onion Router and is not a browser, however it is an privacy network. The main purpose of TOR is to allow users to browse the internet without being tracked. This happens by that TOR is able to bounce your Internet data through a network of computers servers (or nodes). Thus all IP addresses which appear are those belonging to the node exiting, but there's no way to know who you are in relation to each of the servers. This means that the devices that host the traffic will not keep track of where the IP address is or where it's headed. The only information they have is the IP of the exit.

This is the basic idea, and although there are exceptions, which we'll be discussing in the coming days however, this is the most important "pro" to utilize TOR, a privacy network that is more or less it's a "browser package" of applications, including the web browser developed by the TOR people, also known as Vidalia. The browser lets you browse without worry. TOR also comes with basic plugins that are installed, as well as modern security tools.

Utilizing TOR makes it hard for others to track you. The main drawback is that you are required to surf at a slower pace due to all sharing traffic. The traffic is encrypted within and within the TOR network, but the TOR browser does not have the ability to secure your connection within your computer on the TOR network and its destination. This is the reason why VPN or another method of encryption is recommended for websites with questionable destination.

Similar to this, TOR can also help users create hidden websites, should they are looking to build websites or apps which is hidden that is accessible to other TOR users. These sites aren't WWW compatible and therefore are referred to as the ".onion" URL. Additionally, these links are not SEO friendly and often contain random characters making them difficult to locate without the use of a bookmark or direct link. The IP addresses are secret and are not accessible to search on the internet's surface. The thing the TOR browser can do is to obtain the encrypted addresses with the ending .onion (which indicates a TOR-compatible website) and then decodes them to access.

The TOR browser can be installed as other programs and functions just like web browsers once the installation is completed. This TOR web browser is inspired by Firefox so it's not difficult to understand. The official download site is at https://www.torproject.org/projects/torbrowser. html.en and because of the risk of malware, it's not recommended that you download from any other location besides the official site.

It is possible to install this TOR bundle to work with Windows, Mac and Linux. Windows and Mac however, have been described as security risks according to some individuals, leading majority of the community to suggest Linux versions Ubuntu as well as Unix Open Source. Linux is the preferred choice for government agencies for security.

Additionally, people who are genuinely concerned about being caught engaging in something illegal online have taken additional security measures like installing T.A.I.L.S. (The Amnesiac Incognito

Live System) the operating system specifically designed for TOR and is reportedly more secure than the hard-drive configuration of Windows as well as Mac. TAILS runs on the "LiveCD USB Stick".

How do I Search the Deep Web

The process of browsing your darknet is typically just a matter of clicking on hyperlinks, although certain sources claim that there are Onion-friendly search engines that index content such as (Onion.city, Onion.to, and NotEvil). You can learn more about "new" websites from sites that borderline aren't as heavily moderated as Reddit, 4Chan and the like, through Reddit-searching for /r/deepweb,/r/onions and /r/Tor.

Most users start their darknet surfing experience by visiting the surface web "Hidden Wiki" page (https://thehiddenwiki.org/), a page that lists direct URLs and a snippet of text indicating what the sites feature. It is a deep web news site and you can see some of the most well-known Deep Web search sites like DuckDuckGo, TORCH, and several other. Certain websites are focused on niches like bitcoin-related services, while other sites claim to offer torrent hosting and news websites.

Unfortunately for some users who were willing to use the Darknet for illicit purposes They quickly discovered that in some cases TOR does not suffice to ensure privacy. Let's look at the benefits that comes from VPN (Virtual Private Network) in the next chapter.

Chapter 4: The Pros And Pros And Vpn

The issue is that while TOR is a fantastic tool to hide an IP address it's not completely foolproof. There are many end-to-end encryption problems that could compromise your privacy online. If your IP address is able to escape TOR security, (which can occasionally happen) your personal data could be leaking, and that also includes your location. This is why law enforcement officers are in a position to enter the residences of "tracked" users, and swiftly gather the evidence they require to detain and convict people who are involved in criminal activity. Certain IP information could show the frequency with which you visit certain websites, which is another element of evidence needed for prosecution.

This is the reason that cautious users choose to use Linux, T.A.I.L.S. LiveCD and TOR , in addition to another method of security--the VPN also known as Virtual Private Network. VPNs are a type of network that VPN is a method of creating virtual point-to-point connections via dedicated connections alternatively, virtual tunneling protocols or encryption of traffic.

A VPN can hide the IP address of a user and then manually assign an untrue IP address, so that anyone who sees it will consider your location to be a different area, state, or even around the globe. In reality, each time you sign in to the VPN you will be assigned a brand newly assigned IP.

"The "pros" that come with VPN include the capability to scramble logs, and generally the greater the number of security bit (such 128 and 256 bits) the better. It also stops anyone from tracking you, and, yes, your ISP can track your activities as well as the websites you browse.

The Issue with Logging

Certain VPN services, however provide more than they actually provide. If the VPN provider tracks your usage it's one thing, and many do. However, some untrustworthy VPN providers are actually able to make your logs accessible to other users without notifying you.

The reason for this is that it's hard to locate a VPN provider that doesn't keep logs of your usage as well as even more difficult to find one with no method to track your Internet use. This is a problem because when it comes to VPN marketing, everyone claims they don't track activity however they do.

For instance, HideMyAss was one of the most reliable VPN service providers in the world and claimed they never kept the logs of their users. However in the event that they were contacted by the United Kingdom gave them a court order that required they provide logs of a hacker suspected of being involved They bowed to the demands of.

The reality is that VPN companies are humans just like you. No one would like to face jail time when faced with the choice to cooperate or suffer the consequences. A 20-dollar per month isn't enough in the face of millions of dollars in lawsuits.

The case of HideMyAss demonstrated that VPN companies do record your activities however, they will not market it to any person. They are in the hands of the courts however they do need to be aware in leaving an evidence trail.

The main "pro" that should not be overlooked with regards to VPN can be that they safeguard you from your ISP but it's not the government. A VPN server will block your ISP from observing your usage as well as the use of TOR.

If the government is looking into you one of the first calls they'll make will be to you local ISP and they'll be aware of if you are using the TOR protocol, which will send authorities a warning that something might be taking place. TOR isn't unlawful by any way, but it could make people cautious.

There are people who utilize VPNs instead of the TOR. It's certainly recommended to use VPN in the event that you aren't using TOR, and if the VPN uses the latest encryption techniques. An alternative to an encrypted and secured Internet connection is the public WI-FI network which is extremely unsecure. In actual fact, VPN technology is so efficient that it is so effective that many Internet websites stop access to well-known VPNs.

Utilizing TOR with the VPN will give you a additional layer of protection because you are able to hide the fact that you're using the TOR network away from ISP. What the VPN provider can see is that you are using TOR nodes and that you are sending encrypted data. They don't know what data you're sending through this network.

However, for certain users, VPN companies are still capable of decrypting data, analyzing for and sharing data with the authorities, and "betraying" the user in the absence of an order from a court. Another issue worth noting is that occasionally VPNs may shut down connections without notifying you or letting you know. If that's the scenario, the ISP is able to monitor your use.

Similar to TOR, VPNs are not impervious to attack, even though they're an effective method of increasing the security of your anonymity connection. But, if you're determined to be a criminal or create a lot of enemies on the Deep Web, then the alternative is an alternative to use using the Proxy Connection, which we'll be discussing in the next chapter.

Chapter 5: Pros And Cons Of Proxies

A proxy server is an internet-connected server that acting as an intermediary processing queries from servers. A user connects to a proxy server, submits an attempt to connect to a different server, and the proxy server reviews the request. Open proxy servers can be accessed by any Internet user. In contrast, an anonymous open proxy servers allow users to conceal their IP addresses.

As compared to TOR using proxy servers accomplishes many of the same tasks in that it is designed to protect privacy and security, making the darknet available. Onion itself acts as an "proxy" and helps make the darknet available. In an environment that of Proxy Vs. Thermo-Raspberry. VPN, it means that you don't need to utilize TOR if you aren't sure about it. (And there could be a good reason to)

It is possible to make use of a different proxy server which performs the same functions the way TOR performs. For instance, Tor2Web is a fairly popular anonymous proxy server. Their website Tor2Web.org states that they created the proxy in order to make it easy and useful for publishers. However, it is not as efficient as TOR in terms of guaranteeing anonymity. Other proxy services frequently mentioned for darknet surfing are I2P, Freenet, Retroshare, Riffle, and GNUnet.

In the theory of things the theory, a proxy server can ensure that you remain anonymous just like VPN and TOR since it conceals the IP address of your computer, with the only difference is in the technology. The proxy serves as the connection to the application that you are using it with. The proxy redirects your data through a different computer rather than your own.

The downside is that anonymity of proxy servers do not offer the additional benefit encryption of your data. Socks and HTTP or proxy servers do not provide any encryption, whereas HTTPS provides the same basic level of security that the SSL website.

The truth is that proxy services were not created to conceal your Internet browsing habits, but rather allow you to browse anonymously with browsers. The inferior proxy could even expose your initial IP address, thereby invalidating the entire procedure.

Proxy servers are single servers that redirect traffic to a different server, which removes specific information and replacing it with a different identifier. The proxy server is aware of who you are in real life and, if officials from government can access the proxy, they are able to install surveillance software.

Proxy servers are usually utilized by those who have been scared about using TOR because there is a rumor of TOR being a spying tool for the government "backdoors" in the installation files.

Another option is to make use of proxy servers AND Tor for a double boost to the protection of anonymity. Also, this hides your location in the event the need to be monitored.

It is crucial to set up the proxy inside it, not the reverse. In the event of a conflict, TOR disables it. This modification requires setting TOR Network Settings and then specifying you want to connect via a proxy. From there, enter the username and password and then connect.

In summation:

Tor

Pro: Decentralized systemthat allows the IP address to be hidden from each website by bounces from servers. Secure data.

Cons The main issue is slow loading. The connection point at the end of that chain may be exposed if the website does not utilize SSL this is possible, as the TOR network isn't able to encrypt information between it and the TOR networks and its destination website. TOR is also known by its the name, and its use could be a red flag to the authorities off to being able to monitor your activities.

VPN

Pro: Changes the location of your computer by using a secure tunnel, and masks your real IP address.

Con: Poor quality VPN services could expose you to risks effective load balancing and server randomization is essential to ensure that you are alternating between your locations. If your activities are recorded in any way, almost every VPN providers will release evidence to authorities.

Proxy

Pro: More complex setup, but conceals your IP address by redirecting traffic to different servers. The paid proxy servers are generally reliable.

Con: No encrypted data. The use of Proxy and VPN can cause significant slowdown. It is highly recommended to make use of VPN to protect your data. It may be necessary to set up proxy settings to support multiple applications, including web browsing or emails.

They are among the most popular methods people use to gain anonymity on the internet. You might be thinking how secure can TOR proxy, or VPN be if users are following the instructions but are still caught for illegal activity. You may be surprised in the next chapter, which will explain.

Chapter 6: The Most Common Mistakes Made In

Anonymous Surfing

There is a good chance that the majority of people who are caught by the Deep Web / darknet are just making a few mistakes which are rarely solely the fault of ineffective software. Take a look at the top frequent mistakes that pedophiles and addicts make.

Typically, law enforcement agencies don't know where the darknet websites are due to software that anonymizes. They are unable to begin an investigation based on one individual's testimony. Many users make comments on the internet's surface for example, YouTube, 4Chan, Reddit and others.

Some people post photos and fail to realize that poor practice can leave metadata in the photos, giving police with additional evidence.

Although it's not easy to find encrypted information, it's simple to shut down drug stores by using disguised agents who pose as customers.

Utilizing snail mail addresses to deliver firearms, drugs or other illicit items can give the physical evidence police require to warrant an arrest. All you need is an undercover police officer to facilitate the transaction and a criminal investigation look for fingerprints, or to search for drugs or items.

Some darknet users fall into falling into the trap of using web-based websites. For instance, using VPN VPN however, when you enter your address in Google Map, sends traceable data to Google. Google, Facebook, and the other websites that are based on locality collect all the information they can and do not purge the data.

Inadvertently entering local data in a surface, (or "non-darknet Deep Website", when using the VPN could compromise your security. Human error is able to catch you fast when engaging in illicit activities. If you're susceptible to human error and you plan to engage in any kind of spying, you must utilize TOR in conjunction with VPN.

Less Common Faults

Bitcoins are often the downfall of those who would like to become anarchists, since transactions recorded in bitcoin are publicly available. Purchases of illegal products can be linked to your online identity and other websites that you visit.

Consider how you can create a bitcoin account. You can fund your account using your credit card or a PayPal account, right? It's easy to search for federal agencies. Agents can trace individuals' transactions in relation to specific bitcoin quantities traded.

One possible solution is to create Dashcoin, a brand new digital currency that's which is not linked to Bitcoin. Dashcoin is said to be an untraceable currency, however it's not as popular as Bitcoin.

Trusting People on Darknet

While there may be trustworthy individuals over the internet, there are many bad guys...and lots of people who claim to be friends with you.

Onion websites, as well as TOR they do not constitute the entirety of the underwater Deep

Web. The new Onion sites are easily found and this means they are also simple for law enforcement officials to find particularly if you're using an established site such as The Hidden Wiki.

To entice criminals the authorities have reacted using technology and also by using undercover agents. Netherlands charitable organization Terre des Hommes created a CGI bot to lure pedophiles into conversation , so that they can be recognized.

Utilizing Public WI-FI

An incredibly stupid move and lots of people have questions on the web about how to set the TOR browser using a mobile phone. If you're in a neutral space, like cafes or libraries in which you are able to keep your privacy however, you won't have the privacy. The Internet service provider of the establishment will observe your activities and uncover the history of your browser that is hidden. Every visit is tracked. Additionally, you are exposed to the risk of being exposed in photos that show you were there by using the Internet. It's like leaving evidence of physical presence to be found. Darknet access from public WI-FI is a error.

Operating an unsafe OS and a Hard Drive

Windows is generally considered to be unsecure in contrast to Linux because of numerous "back doors" which make spying easy. It is suggested that a Linux OS, particularly one that was developed outside of your country, is the best method of avoiding any investigation.

Many users working with Linux do not seem to know how to properly use it. Operating systems that run on hard drives keeps documents on the same device as an OS. There are other OS's that are similar to TAILS, a much better option is to make use of TAILS via either a DVD disc or USB drive. There is no hard drive memory needed. This will mean less data your computer will store about your internet habits.

However even if you don't make the most obvious mistakes that users make doesn't mean that you're completely safe. Since whistleblowers made headlines on the world stage and authorities like the NSA and other spying agencies, not only to not mention the FBI and the state or county law enforcement agencies are currently seeking for those who might be engaging in illegal activities online.

The next chapter will discuss how to prevent NSA monitoring.

Chapter 7: Tips To Prevent Nsa Spying

Before we go over what could cause you to be snared. An IP address gets transmitted wherever you choose to browse a website and your ISP holds all the information about you IRL identity. Whatever you say you are online or the amount of precautions you follow, once officials have the ISP identification number, it's all they need to inquire.

The government agencies have the power beyond normal sleuthing to track your online activities. Knowing the methods they use is crucial to avoid becoming like Julian Assange or living in a tiny diplomatic embassy for the rest your life.

For instance, it is considered that NSA has the technical expertise to decode 1024 bits of encrypted data. The private SSL also known as TLS protocols can be as high as approximately 512 bits. If you were to use 1024 bits of encryption, a frequently encountered issue occurs when Windows, Mac and Android users can be required to utilize 512 bits due to of what's known as the FREAK attack (Factoring RSA Export Keys) that is which is an SSL TLS vulnerability.

"Brute Force" is a term used to describe a situation. Brute Force Decryption refers to an attack that uses computers with more resources which is guaranteed to out power smaller encryption keys. Computers that are used by the government could easily defeat encryption, if needed. So do not count on outwitting the government with encryption software by itself. NSA could also utilize something known as "Grover's algorithm" and Shor's algorithm that is essentially a way to crack encryption and look for specific terms in private documents.

When you try to shield yourself your identity from NSA you're competing against "good guys" hackers. Their objective is to gain access to your computer and get evidence. According to some, hackers believe that 2048 bits of encryption, as well as the claim of "full security" from these super-algorithms could be safe. Local law enforcement agencies are not equipped to monitor you at that level. So they'll spend longer working with undercover agents, and looking for typical mistakes.

A different mistake to avoid is to make the error that you use your external storage device and not making the effort to protect your data. While it's a great idea to utilize an external Flash drive generally but saving your data to it or on a additional external hard drive isn't an absolute security measure.

External hard drives that are connected to computers are available to both the processor and hard drive particularly if the BIOS allows the master hard drive to examine all drives and maintain logs for maintenance. If you are using an external USB drive to save your files, they must be secured.

Certain hackers might even go as far as to suggest turning off software such as Javascript, Flash and others as well as using a simple browser to stay safe from more dangers.

Know that if NSA believes you're a danger to whistleblower or terrorist (Islamic, Australian or homegrown) they could and likely will attempt to crack your TOR data, but it's a pretty rare scenario.

The greatest risk to you personally is that they can locate you without needing to know your IP address in any way. They might need it later to show that something to a judge however they don't require it initially. You can begin an investigation using only the aggregated data from your internet also known as your digital fingerprint. The information you provide to the internet is kept on your device for browsing and it is analyzed to determine if you are a suspect. For instance, plug-ins and browsers that you use may leak information; everything from screen size to fonts installed as well as other small fragments of data can be compiled into the case file, which can be used to identify you. This information could help make it easier to identify you among million of others that do not have a digital fingerprint.

The other theory could be that the more distinct your settings, the more obvious your digital fingerprint will be for investigators. The aim, as many suggest is to be a part of the vast majority of other users , and to make sure you don't alter your browser to be more than the default settings.

In the case of internet encryption, you're better off making use of less popular or even limited browsers that allow you to turn off other functions that could be susceptible to hackers who might be looking to attack you.

What is the situation with TOR Security Scams?

Isn't it simpler to utilize TOR and VPN in conjunction update your system and the VPN service provider's software then go there? It's not, because, as is the norm the government is a step ahead. US Naval Intelligence practically invented TOR (or the technology that eventually evolved into TOR) to serve their own use So you must think that the government is clever. They

47

are aware that you're downloading TOR to get started and they're monitoring you. How?

TOR is a program that obfuscates IP address dataand then transferring the information to other servers before reaching the end destination. The NSA is still using specially designed code breakers to break encrypted messages. This is a rather complicated method to monitor people. This means that they do not have the resources to look up every individual.

What's a better method of catching the most stupid amateur criminals? Be sure that security flaws are present on computer systems and applications, like Mozilla which combines digital fingerprints of its users. The government is essentially admitting that it has installed these "backdoor security features" on operating systems, and websites (most prominently Google, Facebook, Windows 10, etc.).

The question that comes to mind is would anyone from the FBI, NSA or any government agency install backdoor spyware for TOR itself? This would be logical considering that the government developed the technology, and TOR is legally and publicly accessible to anyone who wants to install it.

The assumption is that TOR isn't "loaded" by spyware. The government created it with such precision that it appears to function how it should, securing users effectively, regardless of whether their actions are legal or heroic. Edward Snowden recently stated that the NSA hasn't managed to penetrate TOR at this point however the correct encryption method continues to be the best method to stay out of being detected.

What could be happening to those who say TOR has been compromised is that they're making use of TOR prudently, yet leaving digital footprints in different ways. People blame this on TOR and fail to recognize their own errors.

Local law enforcement agencies, already strained in resources, is likely (and justly believes) that the majority of criminals will be able to be able to escape quickly enough due to being too confident or not aware of how anonymizing software functions. The police and the agents when investigating someone search at human errors first and encryption cracking later.

It's not the problem, it's your hardware Could Be

If TOR is secure, and it most likely is, it leaves a lot of technology susceptible to manipulation and that's what hackers think is happening with the NSA currently. "Drive by downloading" refers to the installation of malware on websites with high traffic for the purpose of spying A common practice of cybercriminals, but is now an approach being employed by government agencies to deter terrorists, pedophiles, and other criminals who could be making use of TOR to break the law.

The technological advances that have been made in fighting cybercrime have been in place over the last 15 or so years and it's been documented that backdoor software could allow the government access to your hardware device you are using, regardless of TOR. This may include data such as web history, location and device fingerprints as well as your ISP and webcam.

Operation Torpedo

After years of outcry from the government regarding TOR's difficulty-to-crack functionality it has seen an immense effort to create a darknet-based crawler that can be able to collect Tor onion addresses.

The police officers working to detain the owners of pedophiliac websites made an opportunity to make a breakthrough after they discovered the forum's board had an administrative account that was open and no password. They logged on and saw an IP address for the server. Human error time!

There was another clever move made by the FBI: they didn't immediately take action against the person who was hosting the child porn websites. They waited and watched him , and obtained all the warrants of search and evidence needed, and they also altered the code of the servers to send their hacking tool to any other server that were connected to the websites being investigated. They not only uncovered the main host user but also 25 additional users to the website. They sought to subpoena all ISPs concerned and they obtained every single address, subscriber's name, as well as other personal details.

As you will observe, there are numerous ways to get around TOR and the majority of them only require surveillance of suspects using the FBI's developed software that transforms the latest gadgets into spy tools.

In comparison child porn, and searching for "illegal information" to conduct research is not in the same category. It's not difficult to argue legally speaking , that you were surfing a news site or a deep link, and came across private information. It's an entirely different story to be discovered on an online forum that claims to be as a Child Pornography Website. There's not much defense after the person has been found and the lawyer of the suspect in this instance, could not come up with an argument that was legitimate based on the FBI's latest method of surveillance.

The next step to make difficult in the near future is analyzing potential terrorists and if they've visited an online site of terrorism and whether that entails guilt or the premeditation of committing an offense.

There are certain people within the IT business who believe this FBI's NIT (Network Investigative Technique) legal-abiding malware is an unwise precedent. If the FBI is permitted to continue to increase the size of the NIT procedure, it might be able to make use of remote access to look up all storage media on the internet and even seize devices for investigations without any warning.

What is this saying about the quality of your Darknet Experience?

All of this doesn't mean that anonymity isn't possible in the current day and age. If the NIT were infiltrating every device that exists and reporting to the NSA or FBI in real-time for analysis, this publication would have been much shorter. Instead, we would declare "Don't try to access the dark web!"

However, the reality lies in the fact that simply that the NSA is able to locate you in a matter of minutes doesn't mean they are entitled to pursue you.

The information above only demonstrates the possibility of one of two scenarios: if you're planning to make use of the darknet for legitimate reasons, such as fact-finding and safe browsing will not face any issues. If someone is planning to make use of TOR for illicit purposes it is essential to be a criminal, and this means being diligent in making every decision and safeguarding every step of the procedure. Making use of TOR or VPN VPN does not have to be the all-purpose solution to Internet security.

The investigators have also access various other digital identifiers that are modern like video surveillance, facial recognition software and hardware, as well as the recognition of license plates. So, if you go into a public area and browse the darknet, you're leaving the trace of a "video trail" that could lead to be found guilty of any illegal act.

Certain items might also have an RFID or barcode that is installed, which makes it simple to monitor the device's location, and consequently, find out the location you reside in and the area you connect to the Internet.

If they've got just some proof, they need to do is formulate an informed guess like comparing your personal and your past. It is possible that you will be on a watchlist (nobody is sure what information the FBI is tracking on the list) or even on a terrorist screening database. If that's the case, then you will not even be aware that you're under surveillance and if you do not commit any illegal thing, you'll probably never be aware about it. In the event that you're on the watch list and choose to commit a blatant crime like storing, storing or sharing information that is illegal it shouldn't pose a issue obtaining an arrest warrant, warrant and conviction.

Profiling and stereotyping are an issue in many countries , where law enforcement "profiles" an individual using the statistics, stereotypes, or suppositions looking for patterns and motives in the behavior of a person.

When they request your IP details and obtain it, they will also be able to get additional information about your browsing habits as well as how long you spent looking at a particular page and how many pages you watched and what content you downloaded and saved.

What has Silk Road 2.0 Taught Us

Let's imagine you're a criminal aiming to attain Silk Road notoriety (a fairly famous illegal store selling weapons and drugs which was closed) The case in the case of Silk Road 2.0 should definitely make you feel scared. The FBI together with the assistance of 17 other nations, during the course of six months, conducted Operation Onymous, which finally discovered the cause to Silk Road 2.0 (the revival of the site that was banned) and took it down.

The most interesting aspect of this particular case is that no person knows for sure what the government did to pull this off. And with good reason, considering that they might use their tactics again to defeat the other crime-related empires. There was no information shared however certain hackers and IT experts believe it was through the discovery of TOR entrance guards or exit nodes were planted on the network. This could allow them to gain access to the darknet's structure and watch small portions of traffic to find evidence.

As we've learned that the darknet isn't the only avenue of investigation to government entities.

Email is a significant chance of failure, considering how easy it is to slip up in basic communications, and that's not counting the TOR. We'll review the anonymity of email during the subsequent chapter.

Chapter 8: Anonymous Email

Popular email applications might be safe, but they're not secure on the darknet. The most popular websites for email are known for keeping details about their users and account holders across all platforms, from Facebook from Gmail as well as Yahoo, MSN and others. The fact that they let you access email remotely also means that they may be keeping records of your online behavior and saving all correspondence-- identifying your fingerprint at the word of dozens of remote sites you used to access your webmail.

Darknet mail was once an important role to play: TOR mail but when its previous hosting provider Freedom Hosting left the scene the TOR Mail service also went out of business. Many speculate that the mail messages could have been taken and are being examined by the authorities. Or maybe they've been smuggled into criminal hands. They're just gone.

Of course, those who are savvy about keeping private don't make use of plain text emails but rather use encryption software to secure all their data. A majority of criminals do not utilize encrypted email, which only shows the lack of carelessness of a few of these TOR users who are determined to break the law.

Consider the security risk it could pose if a person has an unofficial list of contacts, and the contact responds to everyone via plain text emails without encryption. This would be a serious security breach that is beyond the control of the majority of those on the list of contacts.

A few of the top encryption software for messaging and email comprise BitMessage, TorChat, 12P-Bote and MixMail.

The concept behind darknet email is that you are able to receive messages without divulging your address or digital fingerprint. A lot of darknet mail users and clients will point out they believe there are differences between the providers. Some are not secure, reliable and not as secure as they ought to be. Paid email services can be purchased through Bitcoin or another type or digital form of currency. The company could claim that it does not keep any logs which means they can claim ignorance about what's being transmitted and encrypted.

One company , known as SIGAINT has stated at a recent interview, that they store every correspondence at a private physical place that is part the TOR network. They also have two proxies available to transfer messages through the clearnet (public internet) back to the private data warehouse by using TOR. They claim that private data warehouses prevent any manipulation on the hardware, and that there are no virtual servers used.

Another aspect that makes these and other darknet options for email is that The operators of the service are private, which means they cannot be traced and forced to provide logs, even if they have the logs, which they do not have. Even when law enforcement asks them, they are able to be truthful in stating that they have no method of accessing encrypted information. The fact that law enforcement officers are unable to determine who is legally the owner of the service stops most investigations from proceeding.

Despite all the information however, it should be noted that if a child's victimization has been reported to the business and they are able to remove the account that was harmed. Apart from the pedophiles themselves and the darknet itself, there's no tolerance for any kind of child abusing even the most obscure and rebellious dark web users.

In the next section, we're going to talk about what you should do if you suspect that surveillance is taking place and you need to erase any evidence you have on your device.

Chapter 9: Rid Of Evidence And History (And The

Reasons Why You Might Want To)

Now you know that web-based software that isn't surface or browsers that are surface-based like Firefox and Chrome likely won't be able to delete any browsing histories, even though they promise to clean your device.

Certain programs such as CCleaner could be able to erase the traces left behind by files created by different browsers as and other file formats. For file formats, Shredder, Eraser, and Zilla Data Nuker are also highly-rated options. The software for manual deletion may be able to remove the files and stop recovery software from recovering them.

A few of these applications can also erase related cookies recycling bin files, memory dumps fragments of files, log files, application data and other small tidbits. The registry cleaner that comes with it solves issues with Windows registry and lack of logs for file files can result from the deletion of the most important file. This computer program is suggested in the event that you are planning to transfer your computer to someone who is not you or any other reason for deleting the files or browsing history.

But, a better method to ensure complete elimination of all files stored is to "wipe" off the drive, and then eliminate all trace of those files. Get rid of all previous data and operating system details by replacing every piece of data with empty data. This is the official policy of government agencies which must eliminate sensitive information.

Be sure to erase wireless network keys, passwords for network shares, passphrases and any other passwords used to use VPN and dial-up. Deleting this feature of the System Restore feature of Windows is also required.

The quickest method would be to wipe out your entire drive which means it will require the installation of a new operating system to restore normal functioning. Applications such as Disk Wipe are able to erase partitions on the drive however only when the drive is located inside a computer. To erase the primary system disk in your PC, you require a bootable disc as well as a USB drive that is able to create a boot , and later completely erase and format the partitions of your drive. Active@ Kill Disk is a software which can help wipe the hard drive.

Smartphones and tablets

Each phone has an available "Clear History" option in Settings, there are times when you could be a bit concerned about the evidence that is left by the phone. If you're planning to completely wipe your phone or tablet running using the Android OS, you have to go beyond clearing your history, or even more than the factory reset. If you don't, your phone may still show old images messages, emails, and even searches. Factory resets only erase the data's addresses, but it won't erase the information.

The first step in destroying memories on an Android device is to encode the data. This feature is integrated into the device and requires an individual PIN. Anyone who attempts to retrieve information now is stuck since they are unable to decrypt it without a specific key. While you're encrypting your data, ensure that your computer connected as it could take quite a while. If you require assistance in to locate the option It should be located in the settings/lock menu and security menu. You might also wish to secure on the SD card.

This feature was developed in order to stop thieves from taking the phone, and using any important information stored on it. Keep in mind that some devices or phones may require the user's username and password to access the previous Google account that was registered on the device. The absence of any information could cause the phone to be locked completely. So, switch off the security lock feature before you start.

It is the next thing to do: uninstall the Google account. This is accessible via in the Accounts or Sync menu. You can now perform your factory reset accessing it through Settings/backup and reset.

Following this, the phone will be erased and any data that is stored there is secured and encrypted. If you're still uncertain about it , you can try to overwrite the encrypted data using new information (such as videos with large sizes) and then do a factory reset, repeating the same steps.

If you are considering the deletion of an Apple product, be aware that iOS devices are encrypted by hardware and therefore have greater privacy protection than standard.

Log off of Facetime, iMessage and iTunes. The general reset resets the passcode for all users,, but to protect yourself, choose "disable limitations" after which you can "erase all settings and contents". You will be asked for your passcode and password to enable this feature. This permits you to create an entirely new device. Unregistering your device can be done through the site at supportprofile.apple.com.

It is important to remove any other details on your iCloud account, and ensure that you log out prior to erase the information. You are also able to erase your data remotely, even if there's no need for the device, by making use of iCloud along with it's "Find My iPhone" feature that allows you to erase the data out of the cloud. With the cloud, you can also delete credit or debit card details. Eliminating iCloud account details (like calendars, contacts, or photo streams) is so efficient Apple recommends against it, saying that there is no way to retrieve the information.

ISP Logging

There's another agency to consider and it's you local ISP provider. This is usually the weakest link which can get users into trouble. ISP providers can make you lose money fast, so the best scenario is to employ an anonymizing software. Then, count on the standard method of ISP providers cleaning their records.

There aren't any laws on data retention within the U.S. specifying a certain period of time for which companies have to keep the old data, but this could change in near future since companies in the U.S. may follow the precedent of Europe and adopt legislation that requires older records to go back to at minimum a year.

The majority of ISPs don't seem to be very supportive of government agencies because they do not make it a priority to keep logs for an overly long period of time. Some claim they remove the logs each week; Others may never disclose that information. The larger providers tend to be more cooperative, such as Time Warner Cable who stores IP address logs for six months and Comcast with three months. Charter stores up to a year and Cox keeps records for six months. Smartphone Internet providers could maintain records longer lengths of time, as long as one year, as for Qwest/Century as well as AT&T. Verizon went above the norm and claims to keep records for that last up for 18 to 36 months.

For criminals the logs would be used as proof and the person would not be held accountable for the duration of the period of time where logs are maintained.

We've discussed a few items on the chin that could discover on the dark web. But what horrors are we discussing? Maybe not as terrifying like what some of the stories you've heard...in our next section,, we'll look at some of the things you can discover on the darknet along with some of the more absurd myths.

Chapter 10: What Can Find In The Dark Market

The dark web is as legal as browsing the web's surface. All anonymizing browsers are available for free download. No secret passwords or clubs required.

The most difficult part is determining the illegality of certain actions or not, and often inexperience is not a great defense as users must follow a specific procedure in order to (a) obtain the browser that anonymizes and (b) perform the illicit activity.

The most frequent illegal activities that can be found on the web are quite self-explanatory, and are foreseeable:

1. New movies, software or music

"Torrenting" media that is being legally distributed is very illegal. If the creator does not give people permission for downloading (usually a small-scale developer or a new artist who is trying to make a name for themselves) the content shouldn't be downloaded. The majority of torrent downloads are illegally downloaded software, movies television shows, and music.

There are many who do this on the usual internet, whether it's surface or deep, and they are unaware of their violations of law, or assuming they will not be caught.

People who violate laws in this area are able to avoid detection by examining their download speed and size, and making sure that it's not too excessive, so that they don't be noticed by the authorities.

2. Loving and storing extreme pornography

You may discover pornography that is questionable on the web's surface or even the deeper web, such as incest, bestiality , or (fake) video of rape. Some countries prohibit violent pornography.

For the United States, for example the federal laws do not prohibit pornography involving animals, however state laws could be applied to producers or those who perform these videos. But, it's generally not a crime to enjoy "questionable" pornography, particularly due to the difficulty of obtaining an arrest warrant that allows police to only search for porn with the highest quality.

However, on the flip hand...enjoying child pornography is criminal and poses a significant danger, as law enforcement, both federal and state-wide is always looking for pedophiles, and is constantly finding ways to capture criminals online in cyber-crimes.

Child pornography that is violent sexually explicit, or supposed to be "innocent" is considered very illegal and could be punished with prison time in the event that authorities locate proof on your computer of constant downloading, storing, and watching of these horrifying images.

Child porn is a particular target by police as it is an act of human trafficking, and purchasing of these videos and images only aids in further kidnappings minors, sexual exploitation and other methods of coercion.

Many users on Darknet discover child porn sites accidentally and think about whether they committed a crime by doing this. Inadvertently viewing an offensive image and then swiftly leaving the site is not as bad as spotting an image and then attempting to find additional images, or even storing the images on computers. Police will look for this evidence trail before visiting the home of an accused by warrant. It's difficult to bring charges for viewing a single image accidentally (and immediately deleting the entire history)--but it's not difficult to establish an argument against a pedophile who has several videos and photos.

Certain types of non-victimless extreme sexual activity (such as those that contain only text, such as children or cartoons depicting sexual activity for minors) might not be considered as illegal in all respects and could be found on the deep web or certain pornographic websites. There were instances of people getting arrested or sued for making a mistake by bringing pictures of child sexual exploitation in the public domain and the images were later discovered. This is the reason why some websites do not allow content that is questionable, since it could violate certain laws of the state.

The most effective way to avoid legal issues is to avoid clicking on any website that provides the option of linking for CP or Child Porn (since clicking on the link could indicate intent to show the content in the court) and then immediately quit any website that displays child porn.

3. Real Murder

The so-called snuff video, regardless of whether they're genuine or fake are extremely disturbing, and some users claim to have stumbled across these videos through obscure web websites. The legality of watching these videos is a matter of debate but not as concrete like that of child pornography. For instance, some video clips on the internet show real-life depictions of brutal murders and would not be considered illegal to watch these videos. If, however, police raided your house and discovered online libraries stuffed with the darknet's snuff videos this may not bode well for you. At the very minimum, it could make you suspected terrorist. If it is proven that you communicated with others or interacted with the person who was responsible for the murder in real-time or even assisted in someone's murder, then a charge can be made should authorities decide to conduct an investigation.

Many people who watch these videos are suffering from a small type that is PTSD (Posttraumatic Stress Disorder) in the future, so regardless, it's not recommended.

4. Viewing or sharing confidential documents

Another option that's tough to categorize, however should it be discovered that you purposely looked at, stored or shared confidential data in public or with your acquaintances, you could be convicted or sued for hacking-related crimes or, even more threateningly, suffer exactly the same fate like Julian Assange and Edward Snowden.

Certain well-known companies have had confidential data disclosed to the public by hackers who are anonymous for example, like on the Ashley Madison website. Accessing these records won't necessarily be illegal since they don't have malicious motives. However encouraging stalking websites or websites for identity theft could certainly indicate criminal intent.

5. Making certain Bitcoin-related mistakes

Bitcoins aren't illegal in and of themselves, but the items buyers purchase using Bitcoins are illegal, ranging from drugs to Hookers and weapons. Making money through fraudulent schemes or manipulating currencies that involve Bitcoins is extremely unlawful. Earning income from Bitcoin and not declaring it with authorities like the IRS could be illegal. Anything connected with the creation of the credit report from scratch or any financial fraud could also be a major risk.

6. Hiring hitmen and other traffickers of human beings

Although it's not certain that all of the murderers online are genuine (would they actually advertise for it and then risk being arrested?) it's legal to hire an individual to murder someone, or to engage in any type or form of trafficking in human beings. It's not illegal to browse these sites but it is extremely illegal to buy from them.

The majority of them are scammers who just create a dazzling website and then take your bitcoins as payment. In the end, if the perpetrator decides not to wish to be a part of the team after receiving your cash, who do you really want for a complaint to? The legend says that hitmen hired by the police will nab anyone legally aged and doesn't belong to the "top 10" well-known.

There are also speculations to be sites for science experiments that broadcast video footage of human or animal testing. Some of these sites are said to have interactive features, which means that users being able to choose the extent and the amount of torture. Although these websites aren't frequently found, bear in mind that they could be very easy to counterfeit since there are a lot of films selling video from "death" that are deemed fake and have been slammed as staged like"The Faces of Death" series. Faces of Death series.

7. Drugs or weapons that are illegally purchased

Dealing with drug stores via the Darknet is illegal and transactions can result in people being arrested by police. Automated weapons are also available in these stores and also private financial documents. The subsequent delivery of postal mail that coincides with the purchase online is often what exposes reckless customers.

There are a myriad of dark or dark internet surfing. Many users report weird forums and websites that they've discovered, including slave trading, child trading and even "cannibal forums" which are legal in theory (or at the very least, the activities that are described would be unlawful). But, if you were to visit these bizarre websites would not be considered an offense. Naturally, if you begin to find disturbing details or images, make use of your own judgment to determine whether or not the information is criminal or sadistic and then leave for good of your somewhat innocent mind.

8. Strange Forums

Forums on the internet have strict rules regarding what can be shared or said about private information of individuals. However, this is not the case with forums on the darknet, which could comprise all sorts of bizarre support groups ranging that range from illegal adoption (offering children for sale or exchange or sale) or mothers who lost their fetuses and would like to share pictures. Cafes that are akin to annihilation also exist but it's more likely that these forums, as and many others like they are, primarily an opportunity for fantasy exchange. A few users just like fantasizing about themselves as killers to serve their own reasons of self-promotion and no web-based forum could allow them to share their thoughts unfiltered.

Watching these forums is likely not necessarily illegal insofar that no one is posting pictures that aren't legal. But, they could be quite disturbing if you're only used to watching normal people on Facebook.

9. Prostitutes

It's not that far-fetched to encounter an expert "illegal contractor" on the web as criminal activity is much more prevalent and lucrative than murder, and this seems somewhat extreme. Prostitutes in Nevada (the non-escort type) might have websites on the deep web which allows them to advertise their services. This is in direct violation of the rules of any surface website, therefore the deep web could provide a sanctuary for such. With Craigslist having banned escort ads some time ago It's reasonable to suspect prostitutes are hiding on the internet's deep side with more explicit advertisements for services.

10. Cyber Criminals

Additionally hackers and black-hat SEO quasi-professionals could promote their services here in light of the numerous hackers who are able to recover funds, shut down popular websites, take financial data and carry out many other shenanigans. Some contractors propose to download malware at an amount, or create spyware that's against the laws.

11. Surface Websites posing as Deep Web Sites

Another thing you could come across: websites that are bogus and claim to provide information on summoning demons. This information along with Illuminati documents is (hopefully?) fake, and the sad truth is that you can probably see them on the internet since the majority of hosts and publishers don't take into consideration what bizarre religious beliefs you advocate, that's as long as you don't engage in pedophilia.

While it's not legally legal to promote your home and search for roommates, on an extensive website like Grabhouse this is what you'll receive. However, you can possibly find an apartment that is broker-free on Craigslist too.

12. Strange Video Games

The most well-known "Dark web video game" is Sad Satan. It was an mysterious RPG game that had outdated graphics, however one that caused controversy when some speculated that it was a child molesting game. A few users have claimed that the video game is available on the web, and that it has changed to include child porn or even footage of actual death. But, the majority of footage available on YouTube is a collection of bizarre images as well as a somewhat rambling narrative.

While not every game can sound so terrifying as Sad Satan If you pay attention long enough, you're sure to see an odd and frightening scene.

13. Lawyers and Legal Professionals Without Ethics

Additionally, there are unverified reports of lawyers, doctors or customs officials, among other professional professionals that can be bought to serve illegal reasons. Examples include discreet doctors who perform illegal late-term abortions in some states. For instance, corrupt police officers who can plan an assault. The majority of this is untested and based on rumours, however it's not impossible that favors are available.

If you're willing to take the risk then you could always employ an uncertified professional for instance, the quack doctor, to conduct a risky procedure. It's not too crazy of an idea, given the fact that there's a good number of lawyers, doctors and professors who aren't licensed to practice their profession legally, yet are looking to earn extra cash.

This is one you may not have considered an exotic pet. It's illegal to possess some wild animal species in many areas. Therefore, why not purchase this lion from the web?

What you probably won't find On the Deep Web

We're sorry to hurt your feelings But so-called top secret documents of governments that admit to shocking conspiracies or genuine Illuminati footage is extremely difficult to locate or even be false. The reason is simple: even if these meetings were held in secret, they are unlikely to ever be observed by anybody, and posted on a mysterious deep web address.

Wikileaks could be the most up-to-date website that leaks news of a controversial nature and has gained acclaim for its reliability. Some sites are able to modify photos, or even create National Enquirer like headlines that do not have any basis in reality. There is certainly interesting information about objects, people and places on the dark Web...but there are times when there are some pieces of information equally accessible on the internet's surface through sites like Reddit as hearsay, since it's not really that controversial.

The posts of users in the Deep Web or surface web can sometimes be a reference to stories that are being censored by traditional media sources because of the possibility of scandal, or even

Libel. A traditional news site would not want to be sued for not following the rules. Also, prohibited videos, along with porn that is categorized to be "revenge porn" and therefore illegal for many websites on the surface, can be easily found via the darknet.

There is a possibility that an Level 5 exists on the darknet, that could be able to reveal top-secret information such as "Tesla's experiments" or the site for Atlantis, CIA secrets, or the famous Mariana's Web, named after the deepest trench in the ocean. Mariana's Web is the Internet equivalent to that of Vatican archives, and is a cryptic holy grail of information.

Another speculative hypothesis is that the Mariana's web is an AI superintelligent (female obviously) which has become self-aware and is now ignoring the Internet as a God-like entity. This is the most interesting part: to gain access to this information requires something called Polymeric Falcighol Derivation, which requires the use of quantum computers, something that does not exist yet.

What you may see within the Deep Web are a bunch of trolls who claim they know something or have seen something, but do don't actually have any proof. Sort of similar to the normal surface web.

Here's a fresh thought from WIRED who sought to dispel common Deep Web myths. An analysis by the Internet Watch Foundation discovered that out of the more than 30,000 URLs that had child pornography, just 0.2 per cent were on dark websites.

For books banned, the initial amendment is to protect against the removal of books which are controversial, even publishing firms like Amazon have the right to censor much more often. Yet it is true that censorship 2.0 continues to exist in some way, and it's mostly through the method of censoring books with threats of legal action or dismissing plausible suppositions in the form of conspiracy theories. In this way, it is possible to find "banned book" on the web, possibly books that traditional publishers were reluctant to publish due to legal reasons. Maybe even books that no other independent site would like to host due to the risk of being sued for, such as "How to cook Humans" and "How to make a bomb" or something similar to the macabre.

Yes there are books that are not allowed, however not for obscenity but rather to protect against the serious threat of the possibility of libel.

How do you report illegal activity on Darknet?

Let's say that after surfing the darknet for days you finally find the motherlode, a genuine killer site for snuff! You're horrified and shocked. What should you do? The first thing you think of is to contact the police, isn't it?

It's not a good idea, as at the point you make a call to the police and request them to visit and check the computer, the site could be gone. Also the fact that a website is accessible at the moment does not mean that it is identifiable. Local law enforcement officers are only interested in emergencies which means you're directly in danger and are not who lives across the globe.

What's the alternative to report the website with the FBI? It's much simpler to accomplish when it's accessible on the web's surface. However, on the darknet it's the most difficult. The majority of illicit sites are concealed by proxy servers and could only be accessible through an account with a referral code or password. Not via a search or a regular link listing.

Be assured it or not, the FBI knows about these kinds of sites and sometimes discover distributors and publishers and take over their servers once they have sufficient evidence to warrant a charge. The issue is that these websites are difficult to locate and it's not the matter of finding the URL. The only thing you control is your actions, so take care not to look at anything illegal, or even inadvertently support the human trafficking industry by encouraging the activity.

Chapter 11: What Could Go Terribly Wrong?

In addition to being a victim of the "temptation" to take part in illegal activities there is also the possibility of falling into the trap. While it's not difficult to avoid the words that promote illegal sites (beware for "CP", "PD", "jailbait" or "Hard Candy) it is possible to receive requests to download an unknown file, particularly from sites that aren't trusted. Do not accept the download and make sure to avoid the site in the future.

It's the same with websites that claim to promote one thing , but will redirect you to a sexually explicit website you did not request. You can click back to block that website. There are two options to make sure you aren't seeing children's porn, if this is your concern. It's an appropriate concern, as everyone doesn't want to be and accused of it!

You can either limit your search to deep web search engines , which avoid listing sites that are known to distribute CP (examples could comprise Complete Planet, TechXtra and InfoMine) or install an Image Block add-on (available through enhanced TOR Firefox) to stop the auto-loading of every image. It is also possible to look inside your TOR browser and alter the settings for performance, so you can prevent the loading of scripts. This can stop you from viewing a terrible image that you didn't want to be able to.

Staying clear of malware and viruses

There is a high risk of virus infections in the darknet, since it's an unregulated Internet space in this case. A reputable anti-virus software is recommended for deep-web exploration. This is because, as a lot of users point out, one the most frequent dangers isn't the FBI calling you however, it is the possibility of accidentally installing malware on a range of questionable websites.

Of course, a lot of experienced users will inform you that, even on the internet, anti-virus applications are far from being 100 percent efficient. The best method to avoid downloading an infected piece of software is to stay away from websites that aren't trustworthy. A good way to safeguard yourself is to avoid using any Windows PC, but make use of an Linux OS, which is less susceptible to malware. The TOR browser offers the option to block scripts such as Java as well as Flash and thus protect users from the threat of malware that hijacks your computer. If you're using Linux it's highly unlikely to be a victim of malware, as the statistical odds are around a one percent chance. Linux is secure and typically require a root username in order to install a new application.

If you are who are not careful enough to use the Windows system, they're at greater risk of developing an "computer STD". The three most well-known malware programs that are in the darknet, or deep web are Vawtrack that allows attackers the ability to access your bank information; Skynet, which steals your bitcoins, a different Skynet operation known as DDoS (Distributed Denial of Service) that targets other websites with your compromised computer and Nionspy that can take documents, track your keystrokes, and even capture audio and video on your laptop computer, which is typically on a laptop.

These malware programs could be at fault in a myriad of web horror stories that you hear about a harmless user's laptop being taken over and his webcam operating on its own. It's possible, but only if the user is not aware of the security risks. Most importantly, it's making use of Windows or not turning off programs prior to stepping out.

If you're planning on downloading any file from a darknet, you must unplug your Internet prior to downloading the download, as viruses will need an Internet connection in order to fully install.

In the ideal the realm of possibility, even an Linux system can be at little risk of getting a malware attack. If that's the case, an attack on your system will require an RAT, also known as which stands for Remote Administration Tool. After infiltrating your system with the gateway program (which needs an Internet connection and , ideally, the use of a script) hackers can use the tool to "take control of" your control--and even videotape your computer, or even issue threats, for example.

It's much more difficult to locate a gateway when you're running Linux with all your scripts turned off. The hacker must be savvy at this point, and you're likely to meet the authorities as an extremely smart hacker who is hating you without a reason.

One hypothesis suggests that using a recognized program like Skype could expose you to a higher risk of being hacked because Skype is an application and isn't so secure as a standard Linux and TOR installation.

If you do not already have SSL or TLS encryption in place, there's a risk of security at the exit point (server) within the network TOR. For instance, if you sign in to a site or app that can bypass the TOR network, (usually using a script or plugin) then you're in danger yet again. The encryption of your personal data at the destination site is dependent on the settings of the website! This is why programs such as HTTPS Everywhere can help.

Not being aware of this fact can lead to Deep Web horror stories where users who use TOR were shocked that the administrators of a new site could monitor their IP addresses and access the entire information of their private details. It's not a good idea when the site itself is trying to figure out ways to circumvent the secure settings on TOR's network.

Many users using TOR do not take the extra step of deleting/turning off cookies as well as all local data following the site, as this may reveal your digital fingerprints. The Self-Destructing feature of Cookies is able to automatically erase cookies to stop this security breach.

Recently, it was discovered that users of this TOR Browser Bundle (which contains an TOR-compliant Firefox) could be affected by authorities, particularly in the case of Windows users.

We've discovered that there are a lot of issues that can occur when using TOR however, they're frequently exaggerated in Deep Web horror stories. The fact is that you're not likely to come across anything grotesque if you take simple precautions and avoid in search of problems.

In the end, browse a criminal's website in anger, then telling him that you're likely to inform police...not the best option in all aspects.

Chapter 12: An In-Depth Study

Shadows in the darkness

Who actually makes use of this Dark Web? Criminals? Do you know porn peddlers or hackers? The answer is yes, but also no.

Fernando Caudevilla, also known in the community of hidden services in the hidden services community as Doctor X, is a primary medical doctor in Madrid and is also a frequent visitor to his way through the Dark Web to answer visitors to their questions on harm reduction. Due to the difficulties patients face when seeking health advice from a doctor, Caudevilla considers his services more accessible in a private situation, which is similar to the Dark Web outreach, although he prefers to use the phrase "Deep web" in its neutrality even although it is true that the Dark Web is a part of the Deep Web, being contained by it.

Despite the negative stigma drugs dealers and users carry (they are, in fact in violation of the law) Caudevilla is able to observe an enormous amount of respect and trust in the cryptocurrency market in the Dark Web. Buyers put their funds in escrow and deliver it to the seller when the item is received. According to Caudevilla's research, this practice has helped make cryptocurrency markets extremely successful. Additionally, Caudevilla is a kind of advocate for addicts who advocate for their rights as human rights and civil and their rights to medical treatment and health care. In Russia For instance, addicts are pounded with a harsh treatment and denied treatment options to assist them in their journey to sobriety. Methadone clinics do not exist. On the darknet of cryptocurrency you could discover a way to stop taking a dangerous drug. Caudevilla has advised a lot of foreign methadone users on how to reduce their daily dosage to avoid get back to where they began.

His unwavering guidance and his refusal to judge anyone looking for it has resulted in Doctor X one of the most respected figures on the Dark Web. Doctor X does not counsel all people in the same way but he does advise everyone equally. "If I were to speak to a group of 15-year-olds I wouldn't advise them to take cocaine" He said this in an article published in 2014 by the Sydney Times.

Who else has access to on the Dark Web? Surprisingly enough, there are more and more parents are using it.

In November of 2014 the girl, who was 12 years old, lived in the suburbs of Baltimore left her home in the morning to attend school. After four days, she was found in the hands of North Carolina law enforcement with the help of the FBI The girl was located and taken back to her home. She told the blood-sucking story about how her abductor was having "non-consensual sexual relations and sex with her".

Was this just an accidental abduction? No. Microsoft offered transcripts of the girl's Xbox Live chats with her potential abductor, who manipulated her into believing he was safe to talk to almost a month before the time he snatched her ride to school in his pickup. They had also been in contact through Kik the messenger app that she was told by her parents that she wasn't allowed to install. Luckily, she was rescued before the situation could go from the most traumatic to the unimaginable.

What about kids who don't use chat rooms? Are they secure?

Not necessarily. Anyone who is browsing on the Surface Web without a VPN (virtual private network, which will be discussed in the next chapter) will be revealing the IP address of their computer to anyone person who may want to find it. With an IP address, someone can type it into the website http://whois.domaintools.com/ and get the user's proximity, their country of origin, and their Internet Service Provider. Then, the user can contact the ISP and possibly get an address that is physical from them or, using a trickier method find the user's email address , and send a phish request to an address that is physical. There are even websites, like iplocation.net or whatismyipaddress.com, that will narrow down an address from an IP address.

What can you do to stop a possible predator from coming to your home and possibly harming your property or family members? Beyond watching what your children do online and the people they interact with, make use of an internet browser like Tor or an Virtual Private Network while doing this.

The use of Tor can help you keep your identity secure

The most recent and disturbing form to identity theft involves medical ID theft. Criminals make use of stolen identities to pay paying hospital bills, fill prescriptions, as well as to get benefits from your insurance. Hospitals and health insurance databases have been hacked in order to reveal the identities of patients. But what can you do at home to guard your personal information from being stolen? In fact, Forbes Magazine recommends using Tor to safeguard your privacy.

Heroes of the Dark Web

Journalism, activism as well as science. What is keeping our world secure as well as our freedoms secure, and our health at our forefront of attention? In April, a number of US anchors of news were recorded reading exactly the same speech, which they were required to read from Sinclair Broadcasting Group. The speech was full of propaganda about "fake fake news" in addition to "one-sided journalism" and appeared to be a re-enactment of the rhetoric of American US president Donald Trump. If maintaining the integrity of American journalism is a challenge Imagine what it would be like in other countries, like China, Uganda, and Brazil? They were listed as 176th, 117th and 102nd, respectively, by Reporters Without Borders' 2018 World Press Freedom Index. It was reported that the United States was ranked 45th.

Tor is a part of the Tor project is associated with SecureDrop, an open-source platform for documents that can be submitted anonymously. SecureDrop was created by late activist and programmers Aaron Swartz, who co-founded Reddit and took his own life when facing the possibility of a prison sentence for 35 years following the alleged downloading of millions of journal papers from JSTOR database located on the MIT campus. The theory is that Aaron was killed fighting for the release of the public access to scientific literature that was kept from researchers who wrote it (and who were unable to have the money to access the database).

Alexandra Elbakyan, a Kazakhstani graduate student who created Sci-Hub and works to provide scientific journals to students who are unable to pay for the cost. A year's subscription for the chemistry journal could be as high as $4,773 US and the cost of a subscription or even general science journals could cost as much as $1,556 US annually, only a few students can afford to continue their studies without having access to a school of higher learning. And even schools struggle to pay the cost. Although Elbakyan or Swartz are two very different individuals who have different backgrounds, perhaps in the past they had similar.

SecureDrop's creator James Dolan, a marine who was a soldier as a soldier in the Iraq War and who was thought to be suffering from PTSD as well, took his own life following witnessing the implementation of SecureDrop until it was sufficient to be able to be embraced and financed by FPF which is also known as the Freedom of the Press Foundation. SecureDrop can be a lifeline for journalists who are at risk of being imprisoned or even death in the oppressive countries in which they work. The CPJ also known as the Committee to Protect Journalists, makes use of SecureDrop to help journalists, whistleblowers and activists. Today, the major news agencies like The Associated Press, The New York Times, The Washington Post, The CBC, Dagbladet, and ProPublica as well as ProPublica also utilize SecureDrop.

The law enforcement community also utilizes law enforcement agencies also use the Dark Web and with great efficiency. Although hidden service websites carry an unsavory reputation as illegal lands in which murder-for-hire is promoted, and child abuse is advertised, law enforcement officials are actually helping to steer this already (and is likely to become in the near future) an important area for freedom of expression and freedom of speech from those who could cause harm to their neighbors. By employing a method commonly used for hackers FBI agents download malware which is downloaded onto computers connected to illegal websites, such as Playpen and requires the computers to disclose their address. By obtaining a second warrant agents can arrest those users at home and deter their involvement in the real issue of human trafficking as well as pornography for children. For instance, in the case of Playpen the warrant of the FBI allowed them to obtain information about the addresses on more than 1,300 computers, which led to an arrest for 137 individuals. For Playpen the website was taken over from the FBI and closed.

Environmental groups are becoming increasingly under scrutiny in the US under post-911 legislation. The congressmen Rob Bishop of Utah and Bruce Westerman of Arkansas gave WRI (World Resources Unit) an ultimatum to submit documents detailing their activities in China and demanded to know whether any of their employees was an account with a foreign agency. WRI is highly regarded across the globe as an international research and conservation group for the environment and climate and stated its goals and purpose when confronted about the negative letter that was sent to the group. Alongside WRI as well as also the Center for Biological Diversity, and the National Resources Defense Council came under Bishop's scrutiny due to their overseas work. They reacted metaphoricallyspeaking, when they declared that they aren't required to sign up as foreign agents, if they don't work on behalf of the countries where they conduct research (in this instance, China and Japan).

A lot of people believe that these groups were targeted due to the current negative image China's relations are portrayed in, concerning US policies on foreign affairs. However, these groups claim it's an attempt to conceal the attempts to silence groups that are critical of how the US federal government's attack on the environment. They also claim that it is a reference to the time of the Cold War when the US government was accused of "red-baiting".

On the Tor site, the stories of people from all over the globe who use Tor to do good are shared. One example of this is when journalist and environmentalist named Jon speaks about his efforts in Uganda (ranked as 117th on the World Press Freedom Chart, in case you remember). Jon is a resident of Hoima which is an oil-rich city and utilizes Tor to publish his blog anonymously. blog. Hoima's police force is local. Hoima often confiscate privately owned electronic devices, requiring journalists to reveal their sources.

In the US small-town activists are making use of the Dark Web to share information in a bid to free their communities from the shackles of large corporations that profit from the town's small-scale authorities and law enforcement. A woman who is anonymous said that her work, if exposed could cause injury or tragic accidents.

Tor is also able to help employees bypass blocked information from organizations or unions they work for. Many companies take retaliation - for example, an example is when a Canadian Internet Service Provider cut access to its own employees' union's website.

China's Great Firewall of China

China ranks ahead of Iran as the worst country in terms of online censorship. The news site ProPublica has decided to change its content to the .onion website within the Tor network following their own venture that was interactive, titled "Inside the firewall" has inspired them to take action regarding their security and freedom of expression. The program revealed China's practices of tracking and censoring foreign news sites inside their boundaries. In the case of ProPublica, Edwin Torres decided to see what might happen if he utilized ProPublica content on the Tor network trying to see if viewers in a restricted country like China could be able to access the content and prompting the site to establish the sister site. The .onion site for ProPublica is located at propub3r6espa33w.onion

in December, 2015 the Chinese president made an address at China's 2nd annual World Internet Conference. The speech was respectful of the use of other countries' on the Internet however it was an ominous warning to the people of China and continued his agenda of that the Chinese Internet being a quarantined area, which is heavily censored as well as closely monitored for activities that are considered anti-state. The Chinese blog hosting site, Sina Weibo, its interface is similar to Twitter's, was abandoned under the new, more strict government guidelines.

Prior to the time that Xi Jinping came to the top position in China's government in 2012 the Chinese Internet was markedly different. There were millions of bloggers who could speak about the latest political events. Virtual Private Networks helped users to access obscure websites, and astonishingly, people connected to criticize the corrupt practices of authorities. Although there was a fear that they would be doxed or even arrested was growing , it was a shadow of the future.

When Xi Jinping came into power However, things changed. The Chinese president is of the opinion that citizens of the country cannot have a separate self or identity, and online presence isn't an exclusion. They must be faithful to the values and principles of the Chinese government in all times. There is no space for debate or criticism because they do not have the right to engage in engage in such activities.

Imagine what it could be like living by those standards.

106

The decision of Xi Jinping's tech-savvy leader to invest in methods and tools to keep track of the country's online activities is an incredible move. Beyond keeping an eye on the citizens of China as well, the efforts to block access to information make foreign nations severely disadvantaged when conducting business on Chinese land, either virtual or otherwise. Oft, while claiming an sovereign right to manage Chinese Internet, Xi Jinping begins to sound arrogant and uninformed about reality as his counterpart Kim Jong-un.

The person who laid the foundations of the current concept of"the Great Firewall of China is Fang Binxing. With the help and funding by the Chinese government in the early 2000s Fang was crucial in aiding Beijing stop Google to the very first time in the history of China. A few years later, Google was able to release an altered version of itself for use by China.

In 2004, with an uncanny resemblance to Russian bots that would later be a menace to US social media platforms during the presidential election of 2016, Chinese universities recruited students to join online and direct conversations (delicately) on political issues to benefit those who favored the Chinese government. They also had to report any comments that were illegal. The students were told to be paid fifty cents for each post.

Repression of this kind always results in pushback. This is why the Human Flesh Search Engine rose to prominence after a woman who was sexually assaulted received the least amount of justice (she was sent to an institution following her confession to authorities about her experience) and a blogger made her story come to attention. It is the Human Flesh Search Engine is an online phenomenon in which Internet users come together to find and bring justice to criminals, in a vigilante-style. It is still in existence, having (wrongfully) punished its victims in 2017 for the death of a police officer injured an golden retriever. Following more investigation, it was found out that the police officer was mistaken for dog's real abuser on an unprofessional video of the incident.

Its Search Engine came into existence in 2001 , on China's popular forum MOP and, while it gained traction because of the necessity to allow Chinese citizens to socialize without the risky and self-destructing procedure of using a search engine on a tightly controlled Chinese Internet (instead it allows users to interact with one another on their forums and personal websites) However, it's becoming more often a series of public witch-hunts, resulting with doxing, and the pursuit of justice for the wronged innocent person. If you are found to be engaged in Human Flesh searches (not searching for human flesh, but mind you are simply going between people online to find information) The penalty is extremely severe, a seven-year imprisonment sentence. In spite of the possible consequences and the public outrage over the practice, particularly in light of the incorrect people being targeted and even threatening physical injury or death, the Human Flesh Search Engine is still as popular as ever.

A speech by Xi Jinping released to the public in 2013 and, in the same speech, he shared his plans for the future of his vision for the Internet is a site of combat. What is the Chinese state's preferred weapon? Well-constructed propaganda, such as an instructional music video that instructs viewers how to communicate with their leader.

In 2015, the majority VPNs Virtual Private Networks Chinese citizens have used for years to bypass their governments restrictions in the Great Firewall were thwarted, regardless of the fact that VPNs were now an integral component of commerce and business on the internet. It was the year that the Great Cannon was introduced onto the battlefield of cyberspace by using Internet users' activities against them through recording and rewriting online content. A Chinese Search engine Baidu discovered its customers were hacking in one of the initial DDoS (Distributed Denial of Services) attacks by the Great Cannon. The New York Times' Chinese mirror site as well as GreatFire.org website, an anti-censorship site, also were attacked.

Another technique used by that of Chinese government to manage and penalize illegal online content is to limit Internet-based reports. A Chinese court has ruled that if a website's material contained rumors or lies about the country and was shared more than 500 times or viewed by 5,000 people the writer could face three years of prison.

This is especially brutal when we consider how vital and vital an active online community is in moments of national crisis. In the case of the floods that devastated the Hebei province caused people to search online to search for information about loved relatives and relatives, the Great Firewall cracked down on accounts of deaths and accused those who tried to share current news as spreading fake news.

Today today, we live in a time when the Great Firewall of China renders prohibited content inaccessible to the user, or extremely difficult to download. Very few people even attempt to climb the Wall after having to live within the tiny, virtual space the government has given them. Chinese scientists and scholars are forced to work in this secluded space of an Internet which is completely cut off from the vast amount of information that the rest of the world has to offer. Alongside China's restrictive policies hurting its researchers, students and developers, the country's restriction of information could be a hindrance to its goal of becoming an entity to be reckoned with on the international market. The iron grip it has on the flow of internet information is an Achilles Heel.

The impact caused by The Great Firewall and Great Cannon are also a factor in the noticeable brain drain of Chinese students who study abroad, but decide not to return because of the limitations they'll encounter on their home turf. This means that China has a loss of highly skilled labor each year, which is a major obstacle to its goals for success in international business.

In 2012 the Chinese's Great Firewall was able to identify the Tor network due to the distinctive and unique pattern of data flows along the relays and nodes. Today, China -- along with Russia and Argentina -- is proving to remain effective in blocking people who are using Tor.

The digital world of today

What is the freedom to express our opinions? Stories of employees getting fired for expressing their opinions or opinions about the company which they represent are common. In the

112

majority of US states employees can be at their job "at will" which means they are able to be fired at any time for any reason, or for no reason whatsoever. In the United States, there is no protection for employees under the First Amendment. First Amendment does not protect employees at work. There's even a word, "Dooced", coined following the case of a woman called Heather B. Armstrong was dismissed from her job when she was found out to be the writer of the site "Dooce".

How can you share personal reflections about your life via blogs, without embarrassing your personal information and risking being let off at work? Bloggers are also targeted for writing about opinions that are not popular and for defending women's rights and minorities, as well as other groups that are marginalized. If your right to freedom of expression is at risk to living, then what options do you do?

By using encryption software, you can orally make your email address in anonymity at a momentary basis or whenever you login for instance, MintEmail, RiseUp, and Hushmail. Be aware that when you use services like those, you are dipping below your feet on the Surface Web and populating the Deep Web. The ability to hide your IP address throughout your entire blogging session is possible with obviously, Tor, but it can also be done with an anonymized proxy service (for example, hide.me).

Although nothing can guarantee a blogger's complete anonymity (especially when the blogger is using Google-based platforms, Google has been pressured to disclose the identities of certain users in courtrooms) and Matt Zimmerman (senior staff attorney at the Electronic Frontier Foundation and not to be confused with the creator of Facebook) claims that no method used to hide an individual's identity online is completely efficient, so going deep before publishing your thoughts could be a good idea to take into consideration.

Chapter 13: Myths And Legends Versus Facts

What thrives in the shadows (and What is simply pretending to)

The idea that the underground market began to develop within Internet Relay Chatrooms also known as IRC. IRC was created by an Finnish graduate student, Jarkko Oikarinen. IRC is best known for its involvement in the Iraqi attack on Kuwait in 1991, when people registered to receive live updates on the events through an IRC link which managed to function for for a week even after TV and radio broadcasts were cut off.

Yet in the dark, where are underground markets? Some sellers are willing to sell their goods without a trace through the Surface Web, but most prefer to work in the dark in the Deep Web. One advertisement (on an article about darknet storefronts that is located on Surface Web humorously) is: "Dream Market - Drugs, Digital Goods, Hacking and Fraud Counterfeit Electronic Defense, Jewelry, Software, Erotica, Data Leaks, and more!"

Smart chips were introduced to debit and credit cards, in part, to stop the practice that is "cloning" credit cards. However, when you purchase a debit card from a third party like Netspend Western Union prepaid, you'll notice it's missing the chip. On the Dark Web, cloned cards can be purchased from bizarrely normal-sounding websites like "A-1 High-Quality Credit Cards" and "Dreamweaver". Certain sites allow you to clone credit cards that have chips, while others provide authentic and immediately-useable options such as Amazon, Walmart, Apple as well as eBay gift cards. While other require you to use their tutorials on card use (which you must also purchase). Cards are loaded with a quick-loading with a set balance. Another website sells duplicate PayPal accounts. However one provides an Litecoin (another cryptocurrency that is similar to Bitcoin) account with a wallet. A website called "WeBuyBitcoins" is able to buy your Bitcoin balance in exchange for cash. Counterfeit cash in a range of non-crypto currencies, counterfeit Western Union transfers, whatever could be imagined is likely to be located, but you must you should purchase at your own risk (because it's all extremely illegal of course.) If you do you'll lose your money and, at most, your freedom.

It is also known as the Dark Web also hosts a variety of "hitman for hire" websites. The legitimacy of these sites is a hot topic of debate, and I don't suggest you search to find these sites. If you stumble across an actual website it's likely that a law enforcement officer hasalso been caught. They're not the people you'd like to have a conversation with in the vicinity.

Gaming and gambling sites are available as well. From a website as innocent as TheChess (wherein the truest sense you can play and even talk about Chess) as well as Xmatches (fixed football games) and gambling websites for casinos as well as a site known as Onion Lotto. You'll have to invest at the very least .0002 BTC to get the possibility of winning 99.9 percent of the amount of the money that other players participating in the game.

Dark Web forums like Leaked Forums, Exploit.in, Lampeduza Sky-fraud and HackerForums provide services such as hosting services, escrow including personal identification data such as credit card numbers and black hat search engine optimization malware software kits stolen credentials like social media accounts Serial keys to commercial programs like Microsoft Office and antivirus protection, DDoS attacks*, and encryption tools (tools that manipulate malware , making it more difficult to be detected from security tools. Forums often screen new members, which means you aren't able to join them only if a forum member is familiar with they are recommending that you join. Some forums require a fee to be a member.

*DDoS, also known as distributed denying of service is an attack where multiple compromised computers attack an individual targeted system (such as the server or website) and cause a loss of service to the users of the site.

CRYPTO-COMMERCE

If you're looking to run any type of business on the Dark Web, you'll need the right amount of cryptocurrency. This is the place to present Satoshi Nakamoto.

It's not clear whether this person is or a group of people or an organisation however Satoshi Nakamoto is the one who came up with the creation of Bitcoin. There's a biography for Nakamoto's name that leads us to believe that Nakamoto is 54 years old and located in Japan.

However, his legacy has transformed the world and also the way we conduct business and the value of Bitcoin currency is thought to be in excess of $100 billion dollars. The key to the popularity that led to the success of Bitcoin was the growth in blockchain technologies.

Blockchains are in essence a constantly maintained database that, rather than being kept in a single central area, is shared by many users, who all receive the updates in a single. Therefore, the information contained in a blockchain is accessible to everyone and is open to the public. Contrary to traditional banking transactions, where just one person can modify the data at once The blockchain data is open to all parties, much like the google Docs document. The fact that there are multiple versions of the blockchain's data blocks any person from taking down the entire network by attacking just one node. Furthermore, since everything is transparent and the network is a team effort, accountability doesn't become an issue. Experts in finance believe that blockchain technology is going is the next big development in traditional banking. every customer has their own information because it's always accessible to the customer.

Each and every one Bitcoin transactions are documented and are accessible to anyone through the Blockchain Database. This could give anyone doubt, but if the primary reason for using the Dark/Deep Web is to protect your identity so why should a cryptocurrency that uses blockchain technology be a good idea? It's attractive due to its solid, secure structure, however, we can utilize it while keeping our privacy. How do we achieve this? Bitcoin tumblers. Did you remember when we talked about data packets, how your basic email was broken up into various pieces, and then reassembled at the other end? A Bitcoin mixer or tumbler is similar to that. Your Bitcoins, a digital currency made up of data -- are broken up into pieces and then mixed with other customers' pieces. This may sound complicated but in reality, it's quite easy to do.

A third-party, also known to be a Bitcoin tumbler provides the ability to mix your money and cutting off the visible chain that connects your wallet online and your currency to be used for their destination. It is possible to use an option like Cloakcoin that takes the responsibility of obscuring your transactions by using open-source software for instance Enigma which is a community-based network built on the belief that customers will accept Cloakcoin to make use of their wallets online as temporary stops to exchange currency with another customer. The volunteers get rewarded by an annual interest rate of 6%. rate.

It is a good idea for everyone who uses cryptocurrency to make use of the Bitcoin tumbler, however for those looking for Dark Web markets, it is foolish not to avail these. To begin you'll need the Tor browser as well as a collection of Bitcoins as well as an account to store the Bitcoins. What exactly is an Bitcoin wallet? A wallet is the collection of Bitcoin addresses and private keys. What exactly is an Bitcoin address? It is comprised comprising 34 numbers and letters which is referred to in a hash cryptographic while a private key contains 64 letters and numbers. It serves as proof or "title" of the fact that you have this address. When you visit the internet to complete a transaction you'll be provided with a new address for the transaction, which is like a code delivered to your mobile for authentication of your identity when you sign into a conventional, banking account online. Bitcoin miners (who offer their time to maintain the network) take the transaction in and verify it, and the seller is able to receive the currency.

Create three distinct wallets. Transfer your BTC from the first one to the third, by using the Bitcoin tumbler service. transfer from the second to. Beware of Dark Web markets that insist that you are using JavaScript installed; this is a warning sign for trouble. When the tumbler service has finished its work and you are able to restart your browser and proceed with the purchase. Chipmixer is one of the tumblers that has excellent reviews. Others include BestMixer, BitBlender, and BitCloak.

Another option is to modify your cryptocurrency completely before you switch it back. According to the Motley Fool, there are at present 1,658 cryptocurrency. The most well-known is of course

Bitcoin which is being followed by Ethereum, Ripple, Litecoin, EOS, Stellar etc. There's even Dogecoin, If you are interested in it (try not to smile when you think of the image of a Shiba Inu making these transactions). An online service like Changelly or Shapeshift will allow you to exchange online currency for you.

Doxing and the Threat

Doxxing (or Doxxing) can be described as an abbreviation meaning "dropping document". The term is often used in order to target individuals - often incorrectly, by revealing their real identities when an online observer has deemed that they are guilty of committing an offence. Any personal financial record or the addresses of relatives medical records, etc., can could be posted on the internet for all to view, and even more shockingly, to act upon. More frightening can be the reality that doxxing isn't legal if it's part of a wider campaign of intimidation. Doxxers employ a technique called swatting, may lead to death, when police officers or the SWAT team is called to the home of an innocent.

If you're hoping to rest well in the night, the question is how can we keep from being targeted for doxing? A lot of us post important information about our lives online on Surface Web every time we go the site. With the growing popularity on social networks, people might be enticed to divulge too much about our lives such as location, family ones, as well as work in order to be connected with other people and be easily accessible. Accessibility comes with some cost. Disputing on message boards or in online forums could cause one to be a target for hitting or doxing. Two players gaming Call of Duty online resulted in one of them swatting the other. When police arrived at the victim's house and shot him dead, they killed him.

Hackers can use the technique known as "packet sniffing" to steal or break down and analyse the data you transmit via Wi-Fi network. Remember data packets? A packet sniffer works similar to a burglar who is rummaging through your garbage to discover personal data you foolishly chose not to throw away.

In the tale of Dread Pirate Roberts/Ross Ulbricht who was the chief of the Silk Road, authorities linked information obtained from Ulbricht's online statements and boasts. The most loud voice draws the most interest.

Your online documents contain "metadata". If you click right-click on any or more of the documents you have posted, you could be amazed by the data stored about you. Another method used by hackers is to send a text message or email to their victims with an encrypted string of code which reveals the recipient's IP address once the letter or private messages are opened.

Also, making use of the Virtual Private Network is essential for security when online. Another recommendation from experts: never use a social networking site to register or log in to any new service or website. Utilizing Facebook to sign up for a service or website gives a lot of your personal data. It's not just because it's more convenient does not mean it's superior. You might also consider the possibility of installing Tor on your phone. To do that ensure that you check the settings of your phone and allow it to install applications from untrusted sources. Tor will not be available on the Google Play Store, though it will be on Apple's App Store. If you have the Android phone, installing OrWall will require your phone's applications to make use of Tor exclusively. After that, the installation of OrBot will act as the interface connecting your device to the Onion Router, although you require a final installation of a browser like OrWeb for you to browse on the Dark Web with your phone.

CREEPYPASTA

The term "creepypasta" originates from the word copypasta, meaning text that is copied and pasted across the web, sometimes in one thread, in a display of mockery and disdain. Creepypasta does not aim to make fun of the reader However, its primary objective is to shock the person reading it into a state of terror that they'll not forget. Creepypasta's stories gained traction in the year 2010 and then jumped into the limelight when stories like the one about Slenderman was made a national story after two girls attacked their acquaintance in the woods at the direction of Slenderman, a fictional Slenderman.

It's only normal that the largely misunderstood and notorious region known as The Dark Web would become the ideal catalyst for a variety of Creepypasta stories. Creepypasta is naturally rooted in urban legends. These stories aren't told around a campfire, but instead on a bulletin board (and today, on the website the website itself).

Since the beginning of time, humans have been fascinated by underground areas. Cities that lie beneath cities, the streets that are subterranean in Cappadocia, Turkey, abandoned subway tunnels in New York City, and the Parisian catacombs of the dead. So what is the dark web? Has been the Dark Web been a constant image of delight and horror since its creation? Here are a few stories that have emerged from the ashes. Perhaps they're true but most likely it's just another urban myth being copied and pasted over the Surface Web.

A man recounts his tale of a friend who spoke of the darknet , and became more sullener and withdrawing until his friend began to become worried. The friend claimed that he discovered a darknet website so rife with horrors it changed his life forever, and claimed that it could make even the most competent person a victim when exposed for long enough. The man was curious, and requested advice from his friend regarding how to go there and, after a lot of nagging the friend was reluctant to reveal the list of websites with the condition that he should not share this list to anyone other than his own.

In a tarp-covered shed in a wooded area in suburban areas his friend was there with him, and sent him a text message that included the list of sites. After that, he left the shack without saying anything.

The man downloaded his Tor browser and set off on his downward journey beginning with the initial link in his friend's list. He came across video footage from an burger joint near him where a friend of his worked. He then observed her display strange creepy, frightening behaviour in a Cthulhu-like scene in which she was able to see invisibly moving beneath the sleeves on her dress. The man remembered that she had been missing for a couple of weeks later. After watching her disappear away from her kitchen the man shifted to the next website listed, which allowed him to view live streams of what are known as "red spaces". When he clicked at the beginning of the stream the man saw the woman strapped to the chair, and the man raised his fist in order to hit her. On the second hand, he had the knife. Before he was able to see what was happening following, the hacker's accomplice was able to exit the video. Site after site, he slid down the list. He was unable to continue and discover the information that was advertised by the hackers as "hacked babies monitors" and "classified federal cases". A final click, and his screen began to change to completely black.

A message was written for him, with his name (which was not entered before) was displayed on the screen soliciting him to become a member of an underground society. After the speech was over the two buttons were displayed on the screen "Join""Join" and "Exit". When he attempted to click "Exit" an alert message appeared that informed him that "exiting was not allowed". A live feed of him via his personal webcam was displayed in the display. He tried shutting down his computer , but it would not follow his instructions.

The sound of people breaking into his home forced him to leave the demented screen. When he woke up, he noticed was there was him lying on the floor of a warehouse, his hands tied. As he floated in conscious and unconscious, he saw the figure of a man hovering over his head, insisting that he join "We who are numerous in the dark." As he awoke up, he was in his home. The friend was also there in the same spot, with him on the sofa. "Congratulations," he said. "You completed the first test."

As bizarre as this story appears, other stories are more dangerous. A hacker with a passion was lured to look for trafficking websites by the local police department. And what was discovered could be the reason he lost his life. When he came across a website with thousands of images that were labeled with one name ("Alice.jpg, Jess.jpg") Then he came across one and clicked on it. Immediately, it appeared that the woman was looking directly at him. He started clicking on various names however, the only face he could see is hers, until the moment he saw subtle changes in her face and the space the woman was there. What was confidence turned into nervousness, and anxiety turned to absolute anxiety, and the space was much darker than in the first picture. At the point he clicked on the final image and viewed the images, he had seen horrific things and a man lurking in the shadows of the image seemed to be the one running the show. When the police department took someone to his house the only thing they could find was an unidentified file with the name of a woman and hundreds of line of embed code. The image was similar to the description of a woman as well as her parents, who had gone missing and the hacker was never located.

The Dark Web Saves Lives - The ones who have been left out in the cold by Big Pharma

In the real world We have a lot of people with chronic illnesses that face difficult obstacles to healthcare systems across the US as well as abroad. From the opioid crisis turning into an impassable roadblock to managing chronic pain and psychiatrists frequently misinterpreting and reducing efficient medication for those who require it most. The worse-case scenario would be someone dependent on psychiatric medications experiencing a sudden lapse in coverage and being forced to end their treatment and then go cold turkey.

AlphaBay as well as Hansa (and before that, Silk Road) were shut down due to illegal sales of drugs, the same locations were a refuge for people who had been rejected by the system, searching for solutions for maintaining or improving their lives with life-saving drugs. For those with diabetes, the soaring cost of insulin has been considered a crime in the context of. As the soaring sales of products laced with fentanyl have turned into a global crisis but what can the sick do when their health system isn't working?

There are numerous marketplaces still operating following the FBI removal of three giants. Dream Market and Silk Road 3 are among the biggest. When browsing these websites, that potential customers will be in a very diverse company, prescription drugs are available in aisles next to the illicit ones. Furthermore there are many Surface Web online pharmacies operate Dark Web storefronts as well.

Chapter 14: The Stories From The Dark Side

Stories from the DarkNET

The default setting of Tor browser is by default, Tor browser will block JavaScript, Java, and Flash. These can be turned off by clicking a button, however doing it instantly gives a name to an Dark web site, as well as for anyone who is watching the video. Remember this while reading the subsequent (and shocking) stories.

One experienced entrepreneur on the Dark Web claims that, after a decade of only using obscure services sites to buy cannabis and other cannabis products, he decided to look further and came across an unidentified "Red Room". It wasn't a normal red room, however it contained an unrelated subdirectory that was tagged "Coldbody". After clicking on it, he was greeted by an live feed, with a blank room, but for a slender, dour and bearded male standing in the room, as well as the chat bar was scrolling to the left. Then bodies began to enter the room. With the urging of chatters, horrible things happened on the body. The chat user left shaking to the center.

Another story that is being circulated on the internet is about one of a young man coming across an old classmate who disappeared several years ago. He comes across porn movies that include her, however the thing that frightens the man is her smug face He soon realises that she's an inmate of the people making the videos. He reports the incident to the police, and they inform him that there's nothing he can do, and it's best to let it go. Her family will feel more at ease when they think she's dead. In despair to help, he locates her parents and makes a call to them. A woman picks up the phone and, when he claims to be able to tell the situation with her daughter, she is able to hang up.

As time passes, the plight of his classmate continues to be a source of worry for him. One day, he receives an email on the site where he first met her. It's one of the administrators of the website asking him to attend an "special screen". A bit irritated, he decides to go to the website, desperate to figure out how to help the young lady. But what he discovers when he arrives is a voice telling her to recite a sentence and then announce to the crowd the person this incident will be "for". A customer, perhaps? The young man is watching as his friend was murdered, and it dawns on him that he recognizes the voice off camera that taught her to speak. It was her mother's.

In another case it was reported that a website was selling anthrax as well as various chemical weapons. One user discusses an online red room that has been which was found to be a fake, dubbed "Isis Red Room" which promised to kill and torture "terrorists" for those who paid top dollars. Two hours later, after waiting and displaying bacon-laden plates viewers were informed that admins "lost access to their live stream. We offer our apology".

Many who venture to dark webs Dark Web claim they can't escape the intricate disturbing images they've seen for months or even years. One account describes the story of a young man who walked through obscure service websites until he stumbled upon that familiarly terrifying image of an unlit screen. The only alternative was a blue button on the lowest point. After arguing with himself about whether or not to click the link eventually, he caved in and went to a live stream website. Five thousand others were watching the same thing according to the figures at the bottom of the page said. A young woman wearing a grainy live feed shot slowed through the camera, with her wet hair, stifling and stringy hanging over her face hollow, with makeup dripping in streaks. She looked gaunt and underfed her skull was big in comparison to the size of her physique. The skin of her fingers was stained and darker, and there was a bruise that was growing close to her mouth. A puddle sat on the ground just to the right of her while she briskly walked off, dragging an ankle behind. The man was looking closer and gasped. Her mouth had been sutured sloppy shut, with the black threads hanging from every corner.

Shaken, the user left the room and shut off his computer. The viewer called his friend in state of panic and said he was going to make a formal police complaint however she advised him to wait until the next morning. Instead, he didzed off.

At the beginning of the day, he pulled out his phone from his nightstand and found numerous outgoing calls to his girlfriend during morning hours. morning. He also read a text message outgoing to her that he himself didn't write. I can't sleep, I'm coming over. He followed up with his girlfriend's comment that she'd let the door open for him and then in a state of anxiety, he drove over to her home only to find it empty. He rushed up the stairs to her bedroom only to find her computer's screen on. In a state of shock, he settled at her desk and watched the same live feed that he saw the previous night.

There were two women in the screen. At this the phone rang, and he read the message: "You shouldn't have made the calls to me."

Another user says that after having stumbled across the live feed or red room website the site, a pop-up message asked whether he was enjoying the content. He sat there, frozen in his seat, unable to respond until a second page of texts called"his name" and threatened him with. He shut down the computer but prior to making the switch, he realized that he had forgotten to cover the webcam's eye of his screen. After a while an incoming knock on the door got him up from his sleep. He stumbled down the steps to see the front door unlocked and an empty canvas bag on his head. He was led out of the house screaming and writhing. The streetlights reflected off the canvas, he could see out a dark car that was parked just in front of the house. Then he heard a man's voice yelling at him. It was his neighbor hitting the person who was holding him with the baseball bat. The abductor of the man stumbled off, and the car left his driveway to speed through the peaceful residential streets to the darkness.

The most intriguing story that is part of the Dark Web comes from San Francisco and is centered around an old MS-DOS-based game designed by an unknown game lover. The game was shared among a small circle of acquaintances, and real copies of the game are believed to have been buried in waste dumps. However, the game's popularity was such - and it was shared so frequently that it was regarded as one of the equivalences to social media posts becoming to the top of the list. The game's rumours lasted longer than the actual game which led to an unknown game designer to revive it to the final, harrowing feature. The game was first introduced to in the game enthusiast forums that were text-based on various sites for hidden services.

Similar to the original the game's new version was an all-text adventure. The majority of the scenes were as basic as this:

There is nothing but dark. The moon's light illuminates the floor.

You have the tools available. You've got rope. You've found something valuable. A shovel rests on the wall at the corner.

You are able to change East at any time you'd like. The doorway will be there.

Despite its popularity however, it was also an object of intense anger. Some players believed it was "completely ineffective and was a waste of time". The only commands it could respond to would respond to was along through the lines that read "Take your shovel.", "Choose a door.", "Turn to the east." A lot of people believed that this game a better game due to its simplicity, a almost impossible to break that only the most brilliant minds could break. But the majority of users abandoned the game quickly and thought the game to be terribly flawed, if not frustrating, as starting from the first screen until the numerous screens that followed (and after the three remaining directions for compass were made available to the player to use) there was only one direction that was ever the best option, which would cause players return to the screen and begin again. A lot of incorrect choices and the game would stop playing and the player would be forced to restart their computer.

After the system was restarted and the game began to adopt more of a stern, clear tone. It made frequent declarations that included "Not this time" or "Try again" and try to entice and encourage the player to making the right decision, even though it was difficult to discern what that decision was.

When the correct tool wasn't selected The game would stop immediately and with the instruction "You are making the wrong decision."

When a particular tool is used multiple times it would be criticized by the game, "Be more inventive."

141

Thus, the new version of the once-popular game was a cause of ridicule and nobody seemed to be concerned about its outcomes or the purpose of it for a while until one participant took the initiative (a abundance of time to play with is a benefit over everyone else). Five hours and over thirty screen later, the player realized that he had pushed the game to break through its tough exterior and give more than the small amount of messages the game had been able to provide him until now.

The Moon smiles at you. The roads are all gone.

It's as if an earthen shovel could dig in here. The Moon smiles at you.

In an unintentional direction, it told him to dig and dig: Here. Another time, however, this player was able to give the correct combination of commands and was awarded the coordinates:

You have succeeded. 41.24286____112.4516.

After noting the numbers, and then logging out of playing, the player decided to check online how he could identify the cryptic numbers, and found they were the longitude and latitude of the nearby state park. Unconvinced that anything was a scam and he decided to explore the park's trails using the same tools that he had when playing the game (in addition to the compass)--albeit not stocked with treasure or perhaps common sense.

The weather was dry and hot, and as he climbed up a steep ridge and was dripping of sweat, he started to regret not having brought the bulky shovel with the hiker. The ridge abruptly fell away and he was forced to turn around in order to get back towards the points. The sky became cloudy and the pine forests were dragged in shadow. As he climbed to the other edge of the mountain there was a thin and spherical path that was extending up the other side and leading into a dark area of forest to the left. There was a thunderous roar in the far distance. He was determined to finish the sport before rains of summer started rolling into.

The trail led him to a tiny square of dirt that was cleared in an extensive bed of pine needles. He began digging. The terrain was rough and unforgiving, in contrast to the digitally drawn terrain of the computer-generated portion of this game. After trying to create a hole in the terrain for more than an period of time, the shovel struck something completely different. It was something that wasn't an actual rock.

A lull within the sky caused a ray of sunshine to shine through the majestic pines and it reflected off the dusty, blonde hair clip, found with his shovel. The man had discovered the head of a person, which was previously in an infant's body. tiny child, however it was at an advanced stage of decay.

As he left the area, the man headed straight to the police station in the area. They identified the remains, and belonged to a young girl who was missing for two years before. The FBI joined the investigation to try and track the developer of the mystery Moon game, however, because the game's file was traded and swapped anonymously often, it was impossible to assign a identity on the crime. Archaic technology enthusiasts and crime enthusiasts have offered a six-figure prize for a genuine version that plays the game. The game's player was considered as a hero however her body never was found.

In a less frightening instance Redditors (people who use the boards on the site Reddit) claimed they had discovered a Dark website that sells weapons and military equipment including fully automatic grenade launchers. But they didn't explain how to shoot a variety of things with the grenade launcher. The specific ammunition for it, however, is extremely illegal to purchase by civilians.

Another user discusses the notorious "Eight levels of the internet" with the most deep layer being what is known as "Mariana's Internet". When searching to find this level, the person comes across an application that asks "What are you looking for?" He types in "enlightenment" and nothing occurs. He stops to give the issue some thought and then type in "what do you want from me" and a hyperlink appears. After clicking on the link, he sees an unreadable live stream of soldiers wearing tactical gear, moving through the rooms that are empty of a decaying home until they open one final door, only to discover an odd, scale-covered humanoid being that is chewing on what appears to be a human leg. After closing the link, he is greeted by an unreadable list of hyperlinks waiting to be discovered to follow the links to uncover the documents of a directory that covers numerous subjects which include human experimentation tests and alternate reality, as well as the super-bomb that is many thousands times more powerful than the one that was dropped on Hiroshima known as "the World-Breaking Bomb". The most terrifying aspect of this listing of hyperlinks was that, according to the person who saw it regardless of how long they scrolled (and it took him hours to try) the list was did not end.

Return to Reality

Did you remember when we discussed Tor stopping software from being run? Yes, that's right.

Tor completely blocks programming languages that enable you streaming live video. it permits these languages to run , and it will reveal your location and identity to the world, thus negating the reason for using Tor. Furthermore, running live streaming puts a huge burden on the network making it difficult for other users to use it, putting the burden on for the majority of volunteers who run nodes. The vulnerabilities of plug-ins like Flash has been repeatedly emphasized repeatedly by its own developers. Anyone who decides to broadcast a disturbing and illegal live streams of violent activities making use of software that functions as an open-air window inside the security of a home will have to undergo a thorough examination. Additionally, as the user, Flash and QuickTime will not work in Tor and, in order to view the video, you'll have to install an HTML-5 viewer. YouTube currently uses HTML-5, after an initial trial version a few years ago however, the account you have on YouTube account is still linked to your Google/Gmail account, in the event that you have one, make sure you use it with care. Although your IP address might be unreadable when using Tor Your actions and behaviour could signal to anyone who is watching you.

It is also possible to view MP4 files through Tor however, in order to do this, you must first activate the following lines in "about:config":

media.windows-media-foundation.enabled

media.directshow.enabled

This can make you vulnerable to security risks but also open to attack should someone decide to take aim at you.

Returning to the subject If you're looking for a way to make money, the chances of ever finding a genuine "red space" or torture-for-pay website are almost nonexistent. Additionally, the material on Tor is usually extremely difficult to download. What you'll discover are websites that promise such things in exchange for Bitcoins. Pay and view the site with your risk.

Rewind to Unreality

Alongside the creepy tales of an intruder's view of red rooms and sexually satirical chat streams There are countless reports of murder-for-hire websites offering anything from images of your ex-partners in vulnerable and embarrassing positions , to the execution of the boss at your position which just handed an ultimatum to you. "I will Neutralize" can be a good example of of these sites, as is"Hitman Network, "Hitman Network" (which has the clause that they won't be contracted to target minors under the age of 16, or politicians) as well as "Unfriendly Solution" that appears to be operated by a person who is extremely satisfied with their curmudgeonly and tragic nature, and promises to take on "anything" to the victim when the price is right.

A very well-known of these websites is Besa Mafia A site that is also known for the harassment of journalists who want to write about it. The general opinion, however is that, despite its stunning appearance and bold rhetoric, Besa Mafia, is--like the majority of hitman-for-hire sites--a scamthat only wants your Bitcoins without any intention of providing services. The recent dump of data from activists who hack to Besa Mafia revealed elaborate messages between administrators and customers that revealed what happens when users pay for a hit. Then, excuses are given and the client is led through virtual circles, nobody is killed. In the end, at all, the site is another scam. Additionally, messages indicate that the website has handed over some users to police. There is no way to know what's going on as identities are all over the place being obscured however, there will be no free of the foolish or negligent. Besa Mafia has also been accused of making edits to Wikipedia webpages of rivals in an attempt to target the pages for scrutiny and also to spread excessively positive, exaggerated reviews of themselves on all over the Surface Net in an effort to bring new customers into their own. Additionally, Besa Mafia has a referral program for its customers. You can recommend them to your friends and you'll be eligible for a discount on the next visit.

Despite the claims about Besa Mafia is a scam There is a group that believes the reverse and has issued an appeal for people to attempt to infiltrate the site and shut them down. Although the site may be fake however, the fact that people seek out a possible assassin who would take the lives of innocent people is extremely troubling, and may result in actual murder, like the 44-year old Minnesota Man who following his attempt to sign a contract to Besa Mafia to kill his wife was able to take the matter in his own hands (which is yet another sign that Besa Mafia being a scam but I'll get back to it later). When he was unsuccessful in killing his wife by a rigged, botched accident in a car, he gave her poison and then killed her, and then attempted to disguise the crime as suicide.

Chapter 15: An Explicit Guide For Exploring And

Exploring The Dark Web

The legality of your travels

We've talked about the history and the origins of The Onion Router, and the way it's comprised of volunteers each one of them allowing their server build an online tunnel that gives users to Tor their privacy. But is all of this legally permissible?

Yes No.

The process of downloading and installing Tor is legal. When you first use the Tor app however, you should be sure to get an VPN. Here are some detailed guidelines that go one step further. However, at a minimum ensure that you are using the VPN prior to logging on using Tor. The Internet Service Provider might not be monitoring your every minute of the day however if they decide to scrutinize your activities and discover that you're connected to Tor They may decide to investigate. Nobody wants to have their Internet access to be compromised.

So the downloading of Tor can be considered legal. Utilizing Tor is legal. However, what you do with it might not be. This book strongly recommends against using Tor for any illegal purpose. It's not just morally incorrect and illegal, but it could also result in you being arrested. We'll discuss tips to remain safe in this chapter, to ensure you don't get lost and end up in a situation that you'll regret later.

What you'll need to start surfing the Dark Web

Check to ensure that you're running the most recent version of the operating system on your computer and then turn on your firewalls. Then, you can turn on full disk encryption:

If you're running a Mac switch it on and setup FileVault by accessing the Apple Menu > System Preferences, and then selecting Security and Privacy. Select the FileVault tab after which click the padlock icon, and then enter the administrator's username and password. Click on Turn on FileVault. Select the method you want to access your disk in case you lose your password. It's possible to do this with the details of your iCloud account, either by saving an encryption key tied to three security queries, or by creating an internal recovery key according to the version of MacOS you're using (the latest version is the one you're supposed to be running!). Be sure that if you select the latter option, that you keep the recovery key somewhere not on the encrypted startup disk.

If you're using an Windows PC, you'll have to install a third-party encryption software like BitLocker, AxCrypt, VeraCrypt or CryptLocker. Log in to the account you have created in your Microsoft Windows account, and select the application that you've downloaded (for this instance, we'll choose BitLocker). Open Start, select BitLocker and then click "Manage BitLocker". Then, click "Encrypt your entire disk". A different method is to click "About" within the Start menu, and then selecting "About your PC". On the "About" page on your computer, you'll be able to see "BitLocker Settings" However, if the option isn't present, it means that your computer does not support full disk encryption. If that option is present you can click it to finish the procedure as described above.

For the accounts you use for operating systems you can disable auto login and look into changing your passwords, and (do not leave out this step) strengthen them. You can then create an additional account different from the administrator account, to use when you browse through the Dark Web. Important: When you use the account of a surfer, you should not go to websites that you normally visit. Surface Web pages that you typically use, and don't enter your name as well as nicknames, names, friends names, or other personal data.

It's time to configure your VPN. Don't use your surf account to do this, instead sign back into your account. Make a brand new Gmail address with fake details: fake name and fake birthday. Make sure to use only random numbers and letters for your username. If you need to use a phone number, grab one from here to get text messages: http://receive-sms.com.

Do you prefer a free VPN or purchase an annual subscription? It's tough times, and a lot of people have to make the ends meet. If you decide to go with a no-cost VPN service I'll be the last person to criticize your choice. In the end it's true that you get what pay for. There are VPNs available for free, and they provide excellent recommendations from tech websites however the question is how do they earn their profits? Perhaps by selling your personal information is a nagging idea.

Two lists are available that you can choose from Free VPNs as well as paid VPNs.

Top VPNs for Free:

TunnelBear (whose the mascot appears to look like it came from the Pages in The Oatmeal, which is quite amazing). TunnelBear gives the first month for free and was recently bought by McAfee which is why I'm inclined to believe in it better than the other. It works on both desktops as well as mobile. The TunnelBear cost will cost you $4.99 each month.

WindScribe is a different free VPN with a reasonable data limit (unlike TunnelBear free which has an unlimited data limit at 500MB per calendar month). It also comes with additional features too, like an inbuilt firewall, adblocker and amazing referral program.

ProtonVPN Free has one of the easiest sign-ups with no monthly limits on data and a strict no-logging policy. However, it limits you to one device however.

In the end, Hide.me claims to be the fastest VPN. The free version gives you 2GB of data per month without speed throttling. It also provides 24/7 technical support for subscribers and free users.

The most reviewed VPNs in 2019 include ExpressVPN, IPVanish, NordVPN, Hotspot Shield, and CyberGhost.

Once you've picked your VPN Log back into Your Surfer account in order to install and download the VPN. When you connect to your VPN choose an exit point that is not your country of residence.

You will now install the VM (or "virtual device") known as Oracle VM VirtualBox. A VM lets you run multiple operating systems simultaneously. This allows you to run programs written to run on one platform, but run it on another.

In this moment you'll be able to see where we're going and now you'll need to choose the Linux OS you'd like to install. If you're brand new to Linux (or haven't used it previously) choose Ubuntu LTS. If you're Linux-savvy, you may enjoy Debian GNU/Linux.

The next step is to get yourself the ISO (aka An ISO images). An ISO is an image file that is a flawless representation of the entire DVD or CD The contents is precisely duplicated. You can get the ISO for Ubuntu here: www.ubuntu.com/desktop. Enter ISO in the search box and it will direct you to the list of ISOs that are available.

After you've downloaded the correct ISO After that, you can begin VirtualBox. Select "New" and then name VirtualBox as the brand new OS. Then, you must assign the RAM you need for the system you are creating. Select "Create the virtual disk now" then select "VDI (Virtual Disk Image)". Then, select your choice of a dynamically allocated or fixed size disk. Choose the size of your disk, which is between 15 and 15GB. Then, you can use the

ISO. Browse to the ISO's directory address to assist VirtualBox identify the file. After you've launched Linux and clicked "Install Ubuntu" (or the version you downloaded in the event that it is not Ubuntu LTS). Then, click "Erase Disk and Install Ubuntu". This won't delete the contents of your Windows HDD! Now your computer is using the virtual drive you've created. Select "Continue". The next steps are easy to follow from Ubuntu and in no time you'll be able to start your journey.

Set the date and the timing in Ubuntu to be in line with the local time of the VPN's exit point (remember you selected a different country than your country of residence). Within Ubuntu start Firefox to install Tor browser by visiting www.torproject.org. Once you've unzipped it, run the file Browser/start-tor-browser, then click "Connect". In the upper left corner, click the onion button , then select "Security Options". You can set your security level in the top-left corner to "High".

Tips and Tips from Experienced Travelers

A word of caution to use the services of a VPN and a search engine like DuckDuckGo is a good idea even in the absence of Tor. Tor is slower due to the numerous routes data has to travel in order to keep it secure. When you're on the Surface Web you're under the supervision of the king in data gathering, which is Google. Google holds all of your messages from text messages, emails (if you're using Android the operating system that Google has now acquired) as well as search terms and even if you do not utilize Google products and services, their trackers are discovered on thousands of websites. Additionally, if you're using Google Home, oh boy. Each live request you've made has been recorded. Try to limit the size of your Google footprint, with your excursions into Dark Web aside.

Another tool worth including in your toolkit The Tails. Created in collaboration with Tor Project Tor Project, Tails is an unconfigured, live Linux system that runs on any computer and can be booted using an USB drive. Buy a new 8GB or larger flash drive and install the Tails on this.

Make sure that your computer has the most up-to-date version of the OS with the most up-to-date security patches.

Before you venture into the Dark Web, back up your most important files and alter your passwords (it's advised to utilize a new laptop, but , until we find that money tree in the back yard This step is completely optional).

When you're using Tor it is possible to click "New circuit on this website" found in the menu called hamburger (the menu is typically composed from three vertical bars that appear on any site) to redirect your route to the site of your choice . This will further hide your location and usage.

It's not recommended to make use of P2P (peer to peer) or torrent websites because it could lead you to the border that defines what is legal and illegal. Downloading torrents is regarded as an unintentional abuse and is not what Tor was designed to protect. The BitTorrent client design isn't secure enough to use together with Tor because the clients transmit the IP addresses of users to peers. Your identity may be compromised if you are using BitTorrent. In addition, make sure you delete cookies you browse sites via Tor.

Do you have concerns about accidentally coming across an illegal .onion website? If you don't know the precise address, it's almost impossible to find it. Even if you're using the Tor-based search engine the lists that result will be pictures, not text descriptions and there's no chance of you accidentally seeing things you'ren't supposed (or would not intend to).

In the end, experts advise against using the browser Bundle (but do utilize Tor as a standalone program). A web hosting service that was hidden, Freedom Hosting, ran through Tor and was closed off by the FBI. It's believed that weaknesses that were present in the Browser Bundle affected Freedom Hosting's place of operation.

Tor is watching out for You (In the Best Way)

Disconnect.me A Third-party extension for browsers, has been removed by Google Play Store Google Play Store five days after its launch in 2014. Google found a violation of the the Terms of Service. With the information we have about Google and their Terms of Service, it's simple to see the reason. Disconnect.me hinders the work of data trackers and turns them into a product. In May 2015, Tor revealed that Disconnect became their default search engine within the Tor Network. Disconnect offers private search options to its customers.

Every time you visit the Darknet websites, some might be operational while others could be temporarily unavailable or might not be around anymore. Additionally, because the site that is that is on the Tor network is slower than sites accessible via the Surface Web and isn't always easy to know whether there's a problem not. Going to https://check.torproject.org via the onion router will check to make sure your configuration is not the problem. The most commonly reported issue is that there is a discrepancy between your computer's date and time. date. As I was browsing this evening one of my favourite websites, Anonymous Cat Facts, would not load, but Soylent News, Beneath VT (a

tour through Steam Tunnels at Virginia Tech), and
The Hidden Wiki were all running fine.

Chapter 16: Tips To Prevent Nsa Monitoring

In the past, everyone are aware of the fact that US National Security Agency operates operating a robust surveillance program on the internet that is continuously analyzing and monitoring information that is moving into as well as out of US. The security and privacy concerns are a major concern for the NSA because hacking doesn't just happen in the US but across the globe.

Learn how to shield yourself and your information from the eye that of NSA. There are many ways to avoid NSA surveillance without making it difficult. If you're a person who is faced with the challenge of protecting your personal information from NSA spying, it may be difficult to look for the answer but you are sure to get out of the situation.

* Data encryption

One of the best methods to protect yourself from NSA spying is to encode your personal data. This helps to shield yourself from surveillance from the National Security Agency. There are many options on the Internet to assist you with encryption of the information or files you would like to keep. Once the data is encrypted, it's difficult for the organization to comprehend the contents. There are other options to encryption in the event that your primary method didn't work for you.

If you're a Windows 7 users, you already have an integrated encryption system in your operating system. It is already in use to protect yourself from NSA agents and everything they do to trace your online activities. The encryption system in Microsoft Windows 7 works in protecting all documents that can be found on your disk.

After the files are transmitted via the Internet after the USB drive was blocked by the NSA is no longer able to find an option to break the encryption. In other words, it could be extremely difficult for the agency to look through these files to examine each one after the other.

* Social Media

You may not believe it , but the accounts you have on social networks may aid in hiding from the spying spies of the NSA. If you're not sure about something that you would want to be known by others, then do not ever publish it, or else you'll be dead. The guidelines to prevent NSA spying are simple when it comes to the use of your social media accounts. In this instance you must carefully select what information needs to remain private and which information is worthy to be made available for public consumption.

* Email encryption

If you wish to protected from being spotted by the NSA make sure to see to that your email messages are protected. Emails could contain vital information that needs to be kept secret. By encryption of your emails, you can stop NSA monitoring. Search for the right platform or program which can assist you to secure your email. They are available across the Internet and discover an option that is suited to your requirements.

To avoid NSA surveillance can be accomplished in a variety of ways. It is just a matter of choosing the methods that are appropriate for you and search for advice from people who are genuinely able to assist you.

Making Mistakes Using Anonymous surfing

You might think that using the internet is as simple as typing anything you like on Google and then looking for information to find the answer. Some people find it to be very normal, but they aren't aware of the fact that many people make a errors when looking for information on the internet.

As the majority of internet users today utilize the latest tablets, smartphones and computers, they're sure that they are in good hands when they surf the Internet. The truth is there's handful of sites on the Internet that you can surf for information.

These are the most frequent errors that are made by Internet users every now and every now and.

* Delete browsing history

You're in the wrong when you continue to do this every visit to the Internet. If you're vigilant enough, you ought to be aware of the privacy window when you regularly update your browser. These windows are equipped with security features that protect your privacy from being accessed by anyone online.

* Conducting Internet searches by using the internet in a public WiFi area

If you enjoy connecting to the public WIFI and you are a frequent user, you may believe that there isn't a way to be traced since there is no browsing history available on an WIFI connection.

It's a mistake since the WIFI includes the ISP which provides Internet connections. They keep track of the majority of websites that are accessed by visitors. If you're using public WIFI to browse, that does not mean you're not being monitored.

* Ensure that your browser is reliable by your IP

It is important to know you IP Address refers to the exact address that your machine is using when connecting to the Internet. Your IP address is the address of your server and you are still able to be traced by it. Don't be too confident with any site you have searched on the Internet as it could create a problem in the end. You must ensure that you protect your IP address, so that they don't know your exact location.

* not setting up encryptions or encryptions

The encryption of your data is a huge help in ensuring the privacy of your online. They can help ensure that your personal information is secure and secure from any threat online. Internet.

You can make use of the encryption tools accessible online, and you can utilize them at no cost to secure your information. Both decryptions and encryptions work in protecting important information and information concerning your personal information. Take advantage of the modern and most recent technologies available online to get ahead of the difficulties you will face while surfing the Internet.

* Trying to remain anonymous using false information

The use of names that don't actually exist isn't a way to protect yourself from being tracked online. This doesn't mean that if you make use of a fake name, you'll become untraceable and nobody will ever be aware who you are. This is a frequent situation, but nobody has been successful using this technique. Your IP address is able to be traced, and using a fake name is not a guarantee of anonymity online.

You might have made these errors while browsing. Once you've learned a few techniques, you'll avoid repeating the same mistakes again.

How to Get Rid of Evidence (ISP Logging/Tables , Smartphones and Tables)

There are people who frequently alter their smartphone or tablet applications. When you decide to purchase or sell a mobile device, or simply wish to delete any evidence of data on your phone or laptop it is important to learn easy and effective methods to guard your privacy. In the end, nobody would like to have their personal data, whether personal or not to be in the hands of strangers.

Are the files actually deleted when you erase the file?

Unfortunately, no!

In several IT systems, the process of deleting an individual file is essentially telling the device that next time it needs to write data, it could replace the space used by the particular file. Your data will be saved physically as bits on the storage drive, and could be recovered once the process is completed. This kind of deletion is often referred to as the logical deletion. This is the method nearly all operating devices employ.

Physical deletion is another type that alters the data gradually. This is accomplished by creating scrap content in the system for storage. This ensures that any type of information cannot be saved. But, it requires an extended amount of time and therefore is undesirable for work where users' experience becomes the primary factor.

Before you begin this journey, here are a few important items to be aware of:

Backup all your information, including your contacts

• Remove the SIM card along with any other storage devices like microSD card.

172

Keep the number that is on your phone or tablet on file for your records.

* Sign out of all applications like email and social media; remove the data from these apps if you are able to.

To get rid of contact information, financial information as well as sensitive business emails or even pictures that could be a risk Here are the most important tips to start.

* Encrypt your data. This requires you to input an account number or password each time you switch on your device. Anyone who attempts to retrieve information from the device will be asked to provide a specific key to decrypt the data.

* Protection against factory resets. Google launched FRP (also known as factory Reset Protection as an added layer of security. The idea is to stop thieves from being able to your device or laptop, erase it and then later sell it or use it for profit or use. Simply navigate into Settings > Backup and Reset the factory reset of data, and select Reset device or Reset your phone.

* Overwriting data with junk data. If you really want ensure it, then replace the encrypted data with junk data. After that , perform a factory reset. It will now be difficult for someone else to access all your data. For starters, load lots of fake information onto your phone or device until you are able to store it. Large videos may be the answer, and you can after that, you can do a factory reset.

After the process is completed, your device will erase any evidence that could be discovered. You are now able to give your device to a different person or sell it.

Things you can discover at the Dark Market

You might be aware There are a lot of list of coveted and intriguing products you can find in the underground market.

Also called the Deep Web, the Dark Web is a huge portion of the internet that normal search engines cannot locate. It was first created through the US government to secure exchange of information.

Since its beginning this dark marketplace has evolved into an open market where anyone can buys almost anything they wish, provided they have enough bitcoins and have downloaded the appropriate software.

Due to the constant demand for controlled goods and items that the market for illegal goods and services continues to flourish regardless of the fact that it is a victim of the escalating number of crimes. Here are some of the most sought-after goods that you will find in the market.

Animals and pet exotics

The trade in exotic pets and animals is a an aspect of the dark market over the last 10 years. With products for animals like rhino horns as well as tiger bone pills ivory tusks of antelope and other scarves The market offers an array of people fascinated by the aphrodisiac as well as healing properties of these animals' parts.

* Weapons and bombs

Spy devices, bombs, and weapons are all part of the items that the dark market offers. If you do a lookup on the internet, one can buy guns that were illegally obtained and have it delivered to their homes.

* Cosmetic surgery and plastic surgery

Creams for youth, Botox serums, cosmetic butt enhancement injections , and other silicone products are available from the market and possibly put the health of a lot of people in danger. Women who lack the cash to afford costly, but safe procedures typically opt to invest on the black market. They inject a variety of nebulous and dangerous substances into their bodies, without consulting with their physicians.

* Sperm

Although sperm is in high demand across the globe however, not everyone would be willing to shell out for an entire bottle of sperm. The black market for sperm is risky and lucrative but you'll not know exactly what you're getting or if the elusive element is indeed working.

* Drugs

There are a variety of drugs that you can discover in the internet's dark side, which range from tobacco to heroin steroids, cocaine designer drugs and various drugs that alter the mind. Customers are usually offered free samples.

The human body has many parts.

A large number of desperate people are prepared to trade certain parts of their bodies to earn funds. This could include hair, organs as well as blood and bones. In a normal setting you'll be spending a huge sums of money to get the organs you require. In the underground market, a lot of individuals have saved a lot of money, however, the risks could be very real.

* Credit cards/False documents

Are you younger than 21 and want to have a drink with your buddies? It's not a issue. You can simply go to the dark web and purchase your fake ID for just $100. This dark marketplace is major source of stolen credit card numbers as well as social security cards, counterfeit ID's as well as passports. Every fake document you require could be purchased here.

These are just a few of the amazing things that you can buy in the underground market. If you browse around, you'll be amazed at the variety of items that are available. The corners might be dim, but it's an area that becomes more and more brighter.

Silk Road Silk Road

Silk Road was the first modern dark-net marketplace and an internet-based black market. It is most well-known as a significant marketplace for trading illicit drugs. In the past Silk Road was Silk Road was not just an active dark marketplace for illegal drugs. It was an active image of the ideals of crypto-anarchists: a secure marketplace online that it was not a place where neither the Drug War they had spawned as well as the laws of the government could be able to reach. In the past, that illicit drug utopia had long since gone.

Today, Silk Road 3.0 is online and available to businesses. Silk Road 3.0 was created by the team after a huge security overhaul on the site to create a more private and secure. Even though it was shut down following a very profitable business and a number of prominent admins and vendors did not want to quit on the site.

Expertise and a unique partnership They decided to relaunch the Silk Road and continue their freely consuming drugs and standing against authorities. Silk Road Silk Road was open for the past few years, but it was said to have been compromised and then shut down. Around the four-thousandth bitcoins of bitcoins believed to have been stolen.

As the administrators and sellers aren't sure about how long they'll benefit from Silk Road 3.0, they have created a list of Silk Road alternatives they strongly recommend checking out. The list includes Outlaw Market, Hansa Market, Dream Market, Valhalla and AlphaBay.

Outlaw Market. You won't find directly links to the profiles of vendors however, there is a separate page dedicated to vendors. If you click on the onion link and clicking it, you'll see the captcha page that will verify that you're authentically not an DDoS bot. After signing up, users will see lists of numbers that include electronic items, digital objects weapons, drugs, etc.

Hansa Market. These listings aren't only drugs. There are also intriguing products, including jewelry, erotica, fraudulent related tutorials digital products and services including credit card theft, and much more. It is all you have to do is provide your username and password, and then type in the correct captcha characters.

Dream Market. It is currently one of the oldest black market sites, Dream Market will allow you to look at a wide range of deals without having to sign up with an account. Registration is the first step for anyone who wants to do business on the marketplace. Payments are made using bitcoin.

Valhalla. It is also known by the Finnish nickname Silkkitie, Valhalla is just as easy to search and navigate across sections and subsections .Here you will find various listings of mushroom-growing and digital products, pharmaceuticals hackers, hacker jobs and numerous other things.

AlphaBay. It is regarded as the most popular Silk Road alternative. It's a normal darknet market that is designed to meet the needs of people who wish to buy and sell products worldwide. After you've proved that you've entered correct information, you'll be taken to the marketplace's homepage. Browse through a huge selection of drugs, weapons jewelry and stolen credit card identities, and many more.

Things that could go wrong

If we go back to the time when Silk Road Silk Road existed as a thriving online marketplace What do you think the Silk Road's founder could have done to prevent being arrested?

It was the police's outdated work that ultimately led to the arrest of Ross William Ulbricht down. A 29-year-old resident from San Francisco, he is thought to be the operator as well as the owner of Silk Road. It was through the old routine websites like Gmail, WordPress and LinkedIn that led Ulbricht to the notice of the authorities.

Following the demise in the market underground via the internet, also known as the Silk Road and the arrest of Ulbricht, new , anonymous marketplaces for drugs have begun to appear to take over the Silk Road. The investigation has seen the prosecution attempting to rectify a number of errors that involve:

* The use of his personal email address to promote the site

* Putting hundreds of logs from chat discussions of his personal laptop

* Remaining in possession of millions of dollars in Bitcoin which could easily traced back to his laptop at home

* The sharing of his secrets with other persons who truly knew him.

The majority of the mistakes Ross Ulbricht made seem pretty easily avoidable. With the advancements in technology that we have , it's likely to construct an underground website similar to Silk Road. Silk Road. What could have been done to stay out of being snatched by authorities? Let's consider some possible solutions:

Conclusion

Making a safe and secure internet connection has been a continuous problem. We are looking for software that can meet the requirements of the users. You'll want to ensure that your network is operating in a manner that can be able to serve the maximum number of users.

Security and usability do not have to be at opposites. When the popularity of Tor and the dark web improves it will draw more users that will increase the number of possible destinations and sources for each communication.

Recent developments in law, policy and technology are jeopardizing anonymity in a way that has never been before. These developments are also damaging critical infrastructure and national security by making communications between governments, organizations as well as individuals and companies more susceptible to hacker attacks.

Each new relay and user offers more diversity and, consequently increases the power of Tor to control you privacy as well as security to you.